P9-BYS-954

The Middle Path

The Middle Path

Avoiding Environmental Catastrophe

ERIC LAMBIN

Translated by M. B. DeBevoise

The University of Chicago Press
Chicago and London

ERIC LAMBIN is a professor of geography at the University of Louvain, Louvain-la-Neuve, Belgium.

The University of Chicago Press, Chicago 60637
The University of Chicago Press, Ltd., London
 Published 2007
Printed in the United States of America

16 15 14 13 12 11 10 09 08 07   1 2 3 4 5

ISBN-13: 978-0-226-46853-2 (cloth)
ISBN-10: 0-226-46853-4 (cloth)

Originally published as *La terre sur un fil*

Library of Congress Cataloging-in-Publication Data

Lambin, Eric F.
The middle path: avoiding environmental catastrophe / Eric Lambin; translated by M. B. DeBevoise.
p. cm.
Includes bibliographical references.
ISBN-13: 978-0-226-46853-2 (alk. paper)
ISBN-10: 0-226-46853-4 (alk. paper)
1. Global environmental change. 2. Human ecology. 3. Sustainable development. I. Title.
GE149.L35 2007
304.2'8—dc22   2007002865

∞ The paper used in this publication meets the minimum requirements of the American National Standard for Information Sciences—Permanence of Paper for Printed Library Materials, ANSI Z39.48-1992.

We civilizations now know ourselves mortal.

PAUL VALÉRY

We have liberated ourselves from the environment. Now it is time to liberate the environment itself.

JESSE AUSUBEL

In this world, the optimists have it, not because they are always right, but because they are positive. Even when wrong, they are positive, and that is the way of achievement, correction, improvement, and success. Educated, eyes-open optimism pays; pessimism can offer only the empty consolation of being right.

DAVID LANDES

Contents

Preamble

One day while driving down a street in San Francisco, I saw a homeless man on the median strip holding up a cardboard sign that read: "Why?" The question may have meant many things: Why am I here, poor and helpless? Why am I alive? Why is the world today the way it is? Why should the future of humanity and the world seem so bleak? Perhaps it is because this man was asking himself fundamental questions about human existence that he took to wandering through the streets, resolved to stimulate similar thoughts in the minds of others.

If homeless persons in San Francisco ask themselves such questions, why should not scientists everywhere in the world consider them as well? The practice and teaching of the natural sciences in our universities today hardly encourages basic inquiry of this sort. For more than a hundred years now, scientific method has proceeded mainly by analysis, which is to say by dividing a complex reality into its various component parts and seeking to understand the functioning of each component in isolation from its context. In the nineteenth century, however, synthesis—the putting back together of the pieces of a problem in order to work out its implications and meaning for society—was an important part of scientific practice. This approach is no longer the object of systematic attention, having been left to those few, poets and others, who still concern themselves with fundamental questions.

In this book I seek to answer one such question: Ought we be optimists or pessimists with regard to the future of our planet, and therefore of humanity?

I was born in 1962, the year when the American biologist Rachel Carson denounced the indiscriminate use of pesticides and drew attention to the deleterious effects that substances such as DDT had on the environment, particularly on birds. What students of my generation later read about the environment, and what we learned in college, aroused great anxiety with regard to the problems facing mankind and pushed many of us into the ranks of the pessimists. Yet I went on to have children, became a university professor, and devoted myself to research on the environment—incontestable signs that optimism won out in my case. How, the reader may well ask, have I managed to live with this apparent contradiction?

I wrote this work over the course of the sabbatical year I recently spent as a fellow at the Center for Advanced Study in the Behavioral Sciences in Stanford, California. This was during the first term of George W. Bush's presidency, a time when national and international policies aimed at protecting the environment came under attack. At Stanford University, however, I found myself immersed in a ferment of new ideas that hold out the prospect of institutional and technological innovations that could make it possible to reverse current environmental tendencies—on the condition that the political will to put them into practice can be found. More than ever, then, I find myself torn between dark pessimism and jubilant optimism.

I have tried to reconcile these extremes by bringing together recent developments in the various sciences that touch on the interactions between human activity and the natural environment. The following pages offer a synthesis of ideas that have been current in scientific circles for only a short time. My hope is that it will help readers find their own way between pessimism and optimism.

This work owes a great deal to the research that has been carried out under the auspices of the International Geosphere-Biosphere Programme and the International Human Dimensions Programme on Global Environmental Change, and to the work of Gretchen Daily, Carl Folke, Arnulf Grübler, Paul Hawken, Amory and L. Hunter Lovins, Alexander Mather, John McNeill, and Charles Redman, among other scholars. I am grateful also to the William and Flora Hewlett Foundation for financial support, and to Philippe Mayaux for his comments on the draft manuscript. I am indebted, finally, to Malcolm DeBevoise for his graceful translation of the original French text and a great many helpful editorial suggestions.

> Achieving a relationship with nature is both a science and an art, beyond mere knowledge or mere feeling alone. . . .
>
> **JOHN FOWLES**

Introduction

Observe the tightrope walker: always in motion, searching for the balance that he never quite achieves. Each momentary imbalance is corrected by assuming a new unstable position: only by moving forward can he remain on the tightrope and avoid falling. He constantly modifies his movements, subtly, instantaneously, concentrating his attention on the next hint of danger, and reacting at once by a proportionate adjustment of posture.

Thus, too, the earth: constantly changing, forever far from equilibrium, adjusting by means of subtle mechanisms to every momentary reconfiguration of its physical and biogeochemical state in order to preserve its viability. The great climatic cycles, biological evolution, and the natural changes of the landscape are all part of the earth's repertoire of balancing movements that keep it on its own tightrope.

Imagine now that our planetary tightrope walker carries on his shoulder a small boisterous monkey that fidgets and turns in every direction without having the least notion of the difficulty of the task facing its host. So long as the monkey does not weigh very much, and so long as its movements are not too abrupt, the tightrope walker easily corrects for this additional source of instability. Over the course of the twentieth century, however, the monkey grew in size, until its weight was nearly equal to that of the tightrope walker himself. It acquired the ability to make sudden, highly destabilizing movements. Indeed, its impact on the chemical composition of the atmosphere, the earth's plant cover, the structure of the landscape, and the abundance of animal and vegetable species increased to the point that human activity today exerts as much influence on the planet as natural forces do.

If the monkey continues to jump around as though it were on firm ground, the fall of the tightrope walker—and of the monkey—is inevitable. In that case a worldwide environmental catastrophe will occur, with grave, albeit unpredictable, consequences for humanity. But if the monkey learns to coordinate its movements with those of the tightrope walker, even helping him to anticipate and correct successive moments of disequilibrium, they will continue to advance along the tightrope without accident. The future of the monkey and the tightrope walker therefore depends on the monkey's intelligence.

The Issues

From the beginning, mankind has been influenced by its natural environment and has acted upon it. The process of biological evolution that led to *Homo sapiens* is the result of successive adaptations to environmental conditions, often difficult ones. The earliest forms of social organization and mastery of the first tools were likewise a response to the challenges posed by the environment to our ancestors. The colonization of the planet by the human race was itself possible only by means of a series of adaptations to changing climatic conditions and to resources whose supply and availability varied over time and from one region to another.

The discovery by mankind that it was capable not only of adapting to nature, but also of transforming it, represents an important stage in the history of the planet. Fire was the first tool used to modify the earth's plant cover on a large scale. The progressive domestication of animal and plant species increased the supply of food. Irrigation and drainage made it possible to control the supply of water, freeing agriculture at least to some extent from the vagaries of the weather.

Early on in human history, this new power proved to be a mixed blessing.

The extinction of many species of large mammals in North America ten to twelve thousand years ago may have been caused in part by excessive hunting during the first human colonization of the New World (climatic changes at the end of the last ice age played a role as well). Similarly, some civilizations degraded the land they had placed under cultivation, either through excessive irrigation, which caused a salt layer to form that sterilized the soil (as in the case of Mesopotamia between 2400 and 1700 B.C.), or through excessive harvesting of wood for construction and cooking, which, by stripping away the plant cover, eroded the soil (as in the case of the Indus Valley around 1800 B.C., the loess plateaus of China from an even earlier period, in Ethiopia around 1000 B.C., in Greece around 600 B.C., and a few centuries later in Italy, as well as in the southwestern part of the North American continent, on the lands of the Anasazi and Hohokam societies, about 600–900 A.D. and 1100–1375 A.D., respectively).

Some civilizations, however, were able to avoid environmental degradation. Consider the extraordinary longevity—almost five thousand years—of ancient Egyptian civilization, whose agriculture was well adapted to the ecological conditions that prevailed along the Nile. The Egyptians managed to maintain an equilibrium between the seasonal rise and fall of the river, without disrupting the Nile's deposit of sediments on cultivated land in the floodplain.

The question facing mankind today is whether it will be able to go on improving its standard of living while at the same time maintaining a delicate balance between human activities and the natural world. Recent data from the natural and social sciences, supported by careful observation of current developments, furnish us with a rigorous basis for deciding whether we should be pessimists or optimists concerning the future of our planet, and therefore of humanity—without regard for ideology, blind guesswork, existential anxiety, or regret at the loss of some part of the world's original beauty. Our approach must be multidisciplinary and open to arguments on all sides, for whatever answer we finally arrive at, it cannot help but be a qualified one.

What Do the Optimists Say?

Optimists are convinced that technological progress will make it possible to go on coping indefinitely with the ecological challenges facing humanity. They base their conviction on the extraordinary success of technologies developed during the twentieth century, whose contribution to the health and welfare of mankind no one could have predicted a few centuries ago. They entertain no doubt that this progress will continue in the coming centuries, or that mankind's mastery of nature will continually increase. Optimists are convinced, for example, that biotechnological research sponsored by large private companies will solve all future problems of food supply. They predict that, thanks to new production technologies, the demand for agricultural land will decline and the area occupied by forest will grow in the course of the twenty-first century. Mankind, they confidently assert, will manage to improve the environment while at the same time raising its standard of living. They do not fear unintended or unpredictable consequences, for they conceive of the earth as a robust system within which change is gradual, linear, and uncomplicated by major disruptions.

Optimists furthermore suppose that all change is reversible. If humanity should find itself heading down a dead-end street, it has only to turn around and explore other avenues of development. Underlying this view is the conviction that humanity has the ability, if not also the duty, to dominate nature, whose purpose is to assist mankind's rise toward ever-higher levels of civilization. Natural resources are placed at our disposal, without constraint, obligation, or condition.

Optimists have unlimited faith in the mechanisms of the market, whose self-regulating power they believe is capable of correcting imbalances as they arise and of producing the most efficient possible use of resources. Each person is able to pursue his or her own personal interest, since competition for scarce resources within a market framework leads to a convergence between the individual good and the common good for both present and future generations: when a resource becomes less plentiful, its price rises, prompting users to search for substitutes before the resource is irremediably exhausted or degraded. Optimists are convinced that progress in the environmental domain is spontaneous and owes nothing to guidance by national or international bodies. They regard governmental intervention as a source of interference with the proper functioning of markets. Moreover, if a new technology poses risks for health or the environment, these risks are probably less serious than the ones that the technology makes it possible to avoid.

Optimists are fond of reminding pessimists that the population of the world today considerably exceeds the alarmist predictions of the past one hundred and fifty years, and that imminent famines (regularly forecast to occur until the 1970s) have failed to materialize: between 1960 and 1995, world food production almost doubled (197 percent) while total population increased by 188 percent. And if famines persist in certain parts of the world, they are due mainly to civil wars and the disastrous management of agriculture. Optimists find confirmation of their views in the continual decrease, for a century now, in the extraction cost and market price of many natural resources—proof, they contend, that these resources are becoming ever more abundant.

Looking back upon the extraordinary technological and economic development of the twentieth century, and its rapid diffusion from Western countries to the rest of the world, optimists are persuaded that the surest way to protect the environment in the future is to promote rapid economic growth by reducing the intervention of the state in the management of natural resources, thus freeing markets to a still greater degree. Some even advocate the privatization of resources such as water and wild fauna. Private ownership of these environmental goods, they maintain, would reveal their true value by subjecting them to market forces: if this value is high, the market will react by protecting them and by developing substitutes. Optimists note with satisfaction that every generation commits the error of underestimating the number of new ideas yet to be conceived.

The first great optimist was a Frenchman in the late eighteenth century, the Marquis de Condorcet, who firmly believed in the perfectibility of human nature. Imbued with the spirit of the Enlightenment, Condorcet placed his faith in the capacity of the human mind to overcome any and all obstacles that stand in the way of the progress of mankind.

What Do the Pessimists Say?

Pessimists are persuaded that there exist inherent limits to technological progress and that, because it cannot keep on growing at its current pace indefinitely, technology will never be able to free mankind entirely from its fundamental dependence on natural resources. It is therefore necessary in their view to preserve nature's capacity to generate those goods and services that are indispensable for human development. They believe, moreover, that technological development obeys the law of diminishing returns: discovering new technologies that promise to substantially increase the supply of natural resources will become more and more difficult, with each discovery being more costly than the last, while yielding ever-smaller gains in productivity.

Pessimists are more concerned with changes in the stock of natural resources than with changes in the amount of goods produced using these resources. Whereas optimists see the continuous increase in the exploitation of natural resources as proof that these resources will go on being ever more abundant and ever less expensive, pessimists fear that an uncontrolled increase in the rate of extraction draws humanity nearer to the moment when these stocks will be exhausted, not only in the case of nonrenewable resources (such as oil) but also of renewable resources where the rate of exploitation exceeds the rate of natural regeneration.

Pessimists regard nature as a vast complex adaptive system whose evolution is not bound to follow a progressive and continuous trajectory. In their view, the possibility that surprises may occur, some with potentially catastrophic consequences, cannot be excluded. Consider, for example, the seasonal thinning of the ozone layer above the Antarctic. If bromofluorocarbons had been used, rather than chlorofluorocarbons, as refrigerant gases by industry from the 1930s onward, human health would have paid a heavy toll, for the destructive power of bromine on the ozone of the stratosphere is one hundred times greater than that of chlorine. It was only by chance—the relative properties of the two gases having been unknown when chlorofluorocarbons were first introduced for industrial purposes—that mankind escaped a major ecological catastrophe. The Dutch chemist Paul Crutzen, a Nobel laureate, points out that the use of bromofluorocarbons would have led not merely to a seasonal reduction in the concentration of stratospheric ozone (which filters out the dangerous ultraviolet rays of the sun) over an uninhabited part of the earth (the Antarctic), but to a permanent and worldwide reduction, with far-reaching consequences in the form of skin cancers and cataracts. The question arises whether such luck can reasonably be expected to continue indefinitely in the future.

Pessimists consider that certain changes caused by mankind to natural systems are irreversible. Humanity, they argue, cannot allow itself the luxury of conducting possibly fatal experiments with the earth, for it is the only one we have. Poor environmental management is liable to so profoundly degrade natural resources that the earth's ability to provide services essential to human welfare would be permanently impaired. What would happen, for example, if excessive use of pesticides were to diminish insect populations to the point that the natural pollination of crops and fruit trees no longer occurred? Few countries in the world would be able to imitate the example of China, where peasants in Maoxian County pollinate every flower of every apple tree by hand. Pessimists therefore advocate what is known as the precautionary principle: if a risk of adverse effects on human health or the environment can plausibly be demonstrated, controls should be instituted even if the relevant cause-and-effect relationships are not fully understood.

Pessimists see progress as consisting not in economic growth and technological change, but rather in social development. In their view, humanity has a duty to promote cooperation among individuals, in order to maximize the common good, and to place technology in the service of humanity (rather than the other way around). It has an obligation to encourage innovation that favors development without degrading the natural environment and to regulate the behavior of the market in the interest of protecting the common good over the long term.

Pessimists point with particular alarm to the collapse of ancient civilizations that degraded their natural resources to the point that a critical threshold was irreversibly crossed, without anyone having perceived the scale or the immediacy of the danger in these societies beforehand, despite their often considerable technological sophistication.

The first great pessimist was the Reverend Thomas Malthus, who at the end of the eighteenth century predicted that famine, disease, and war would follow upon an increase in human population that outpaced the increase in food production. Pessimists are convinced that if the ecological disasters predicted over the last few centuries have not occurred, it is owing to their warnings and to advances in environmental science, which have led to better management of natural resources. The accuracy of such predictions is to be measured, then, by a society's ability to prevent them from coming true.

Questions for the Future

Must we therefore listen to Cassandra, or can we go on believing the myth of the horn of plenty? The current environmental debate, in opposing the descendants of Malthus to those of Condorcet, bears upon three distinct issues: the magnitude of the changes to which the earth has been subjected; the causes of these changes; and the vulnerability of human societies as a result of these changes.

The first question is one of measurement and requires that objective data be assembled concerning physical, biological, and chemical changes. Substantial progress has been made in this connection. Although uncertainty persists with regard to the exact extent of certain changes (desertification, for example), only a biased and intellectually dishonest interpretation of the factual record would lead one to deny the existence of major environmental changes over the past three hundred years.

The second question is whether observed changes are to be attributed to human or natural factors, or to a combination of the two. The justice of ascribing most recent environmental changes to human activity, at least in part, can no longer be seriously disputed. However, the precise mechanisms that lead a society to degrade or improve its environment are not well understood in all cases, and simplistic notions continue to enjoy currency.

The third question, having to do with the vulnerability of human societies in the face of environmental change, is still more complicated than the two preceding ones, in part because it lends itself to arguments that are based on a great many assumptions and influenced by highly subjective assessments of the place of mankind in the world. These arguments are sometimes extremist, often ideologically motivated, and rarely supported by reliable scientific information. Much of the evidence adduced is anecdotal in character, fails to give an adequate picture of the global situation, and covers insufficiently long periods of time. Yet objective information is readily available. Since the early 1970s, for example, earth observation satellites have furnished data about the state of vegetation, soils, coastal areas, the atmosphere, oceans, and glaciers that, in combination with other findings, make it possible to detect specific environmental tendencies in each geographic zone.

The current debate must go beyond a simple discussion of recent environmental trends, however, and seek to analyze their implications for the future of humanity. Ultimately, the issue is whether mankind can continue pursuing its current mode of development indefinitely. The various parties to the debate defend radically opposed points of view depending on the place they occupy in society, whether they see themselves as winners or losers, and whether the consequences are short term or long term. It is hardly surprising, for example, that representatives of the petroleum industry should not hold the same view of global warming—caused in part by emissions of carbon dioxide that their own activities generate—than an inhabitant of an island in the Pacific Ocean who fears seeing his home submerged as a consequence of a rise in the level of the sea. Similarly, the perspective of business executives and political leaders who pursue short-term economic and social agendas is apt to differ from that of environmental activists who are committed to the defense of certain threatened animal species, indigenous cultures, or the welfare of future generations.

In the pages that follow, I shall try to examine the main issues of this debate in a clear and evenhanded manner. I begin by presenting a summary of data regarding the impact of human activity on the natural environment in recent centuries that cannot easily be challenged. I then go on, in the second chapter, to review the ways in which interactions between human activity and the environment have been conceived in the history of Western scientific thought.

The third chapter presents a few simple models that make it possible to identify the key factors shaping the environmental impact of human societies. The fourth chapter seeks to go beyond the somewhat unrealistic simplicity of these models by inquiring into the fundamental causes of the environmental changes caused by mankind. The fifth chapter looks at two contrasting historical situations: an ancient society that collapsed as a result of degrading its environment (the Maya of the classic period) and several modern societies that have succeeded in improving their natural environment after a period of rapid environmental degradation (certain European countries during the eighteenth and nineteenth centuries). The sixth chapter considers the desertification of Sahelian Africa in order to illustrate the importance of having a thorough understanding of the nature of environmental changes and the need to be able to properly characterize them in all their complexity. The final chapter reviews contemporary developments in Western societies with regard to the direction of technological, institutional, and cultural changes that suggest a reversal of current environmental tendencies is not impossible.

By way of conclusion, I lay out my own views regarding the mode of economic development that will need to be followed if a major ecological crisis is to be avoided, and I propose a number of things that individuals can do in their daily lives that would make a difference.

A Mixed Outlook

This work is the bearer of both good and bad news. The bad news is that mankind has so profoundly transformed the earth that nature's ability to produce the goods and services that are essential for human life (fertile soils protected against erosion; potable water; clean air; a diversity of genetic resources permitting the development of pharmaceutical and agricultural products; ample supplies of fish, game, and fuel; protection against flood; climatic stability; the natural beauty that nourishes the spiritual, aesthetic, and symbolic life of human societies; and so on) is gravely threatened—to the point that the response of the natural environment to further escalation of human pressures upon it has become unpredictable. Humanity is no longer immune to bad environmental surprises. The pessimists are right, then, in insisting that humanity is following a trajectory of development today that cannot long be sustained.

The good news is that mankind has given proof throughout its history of a great capacity to adapt to changes in its environment. Human creativity is, it is true, a powerful source of innovation. By modifying technologies, institutions, and attitudes toward nature, it holds out hope that the pressures exerted by human activity on the environment can be eased before it is too late. The optimists are therefore also right in claiming that innovation makes it possible to prevent changes in the natural environment from threatening the future of humanity.

But innovation on the scale necessary to redress the balance of human activity and natural processes will not occur spontaneously. It will require major social and economic reforms over a period of several decades.

In the past, human adaptation to natural changes occurred only when societies perceived no other choice. Today, given the global extent of environmental change, the inertia of natural and social systems, and the growing complexity of the international economic system, human activity must be adjusted in a deliberate and planned manner before critical thresholds are reached. A general awareness of the risks we face, leading to a rapid and effective response by means of appropriate policies, is indispensable if we are to regain a sustainable course of development. Looking to the long term, we can be optimistic only on the condition that pessimism helps us in the short term to change the world in which we presently live.

Without the willow, how can one know the
beauty of the wind?
LAO SHE

1 The Acceleration of Planetary Change

Modern societies are confronted with the need to strike a balance between adapting to nature—a generous provider, but one that operates in complex ways—and transforming it in order to increase its supply of goods and services for human purposes. For more than two centuries now, our relations with nature have been profoundly modified in both quantitative and qualitative terms: the scale, intensity, and rate of the transformation of nature by mankind have increased by several orders of magnitude since the Industrial Revolution. This transformation accelerated considerably during the twentieth century. We find ourselves living today on a planet where few natural processes escape the influence of mankind.

In 1873, the Italian geologist Antonio Stoppani spoke of human activity as a "new telluric force that, by its power and its universal character, may be compared to the greatest forces of nature." In 2002, the Dutch chemist and Nobel laureate Paul Crutzen declared that the earth had entered a new geological epoch, Anthropocene, characterized by the growing impact of human activity upon the terrestrial system. On this view, the preceding geological epoch, Holocene, came to an end in the late eighteenth century with the initial increase in the global concentration of carbon dioxide and methane in the atmosphere as a consequence of industrial activity. The transition to the Anthropocene coincided, then, with the invention of the steam engine, in 1784, by James Watt.

The Variety of Environmental Change

There is not just one environmental problem, but many environmental problems. They differ in their physical character, occur at different times on different spatial scales, have various causes, and affect different segments of society. The major problems may be grouped together in five related categories:

1. The degradation of land resources, which affects the environment as a factor of production. This category includes the erosion of soils and desertification, which have a major impact on the potential for agricultural production.
2. The use of the environment as a dumping ground, whether in the form of solid waste in garbage dumps, liquid runoff into rivers and underground aquifers, or toxic gas emissions into the atmosphere.
3. The problems associated with the management of planetary resources that are shared by the whole of humanity and that are not the property of private entities or of nations. Under this category come the management of the oceans and the human contribution to climate change.
4. The degradation of nature, particularly of its biological and genetic resources. The crisis of biodiversity is the central problem falling under this head, which also includes modification of the aesthetic qualities of the earth's landscapes.
5. The exploitation of nonrenewable resources, whether in the form of mineral deposits or fossil fuels. When development is based on running down natural capital rather than using revenues generated by this capital, the practice of mortgaging the future raises issues of fairness toward succeeding generations.

To a large extent these categories overlap. A change in the chemical composition of the atmosphere on a global scale, as a result of pollution caused by the burning of fossil fuels, alters the climate, and therefore the productivity of the land as well. Climatic changes combined with pollution and the replacement of natural vegetation by agriculture or human infrastructure diminish biological diversity in turn.

Some environmental changes become worldwide problems by reason of their systematic character, which is to say that they affect a system that functions on a global scale. Atmospheric pollution in industrial regions, for example, strengthens the greenhouse effect and modifies the planet's climatic system. Most environmental changes pose problems in a cumulative manner: small changes repeated in numerous places lead to a worldwide crisis. The extinction of animal populations in one region of the world would be less of a problem were certain species not threatened with extinction everywhere they are found, in which case their loss becomes irreversible.

In other cases, a local environmental change becomes a global problem due to the simple fact of its extent. The explosion of a reactor at the Chernobyl nuclear power station in Ukraine in 1986 not only affected a large part of Europe on account of the radioactive cloud that it generated; it also revealed the vulnerability of humanity in situations where the loss of control over a sophisticated technology has disastrous consequences, no matter that the probability of such events occurring is very small.

The distinction between environmental problems caused by human activity, such as pollution and deforestation, and natural catastrophes that affect human societies, such as earthquakes, droughts, and floods, is now becoming blurred. On the one hand, certain "natural" catastrophes are indirectly caused by human activity: climatic changes triggered by atmospheric pollution are likely to increase the frequency of extreme events, such as droughts, tornados, and floods; human alteration of the earth's vegetation makes forest fires more frequent and more destructive. On the other hand, demographic and social changes lead an ever-higher fraction of the world population to inhabit places exposed to natural hazards: forty of the fifty fastest-growing cities in the world are located in areas where the risk of earthquake is high (Mexico City, Istanbul, and Jakarta, for example); in Bangladesh poor families having no other alternative live on land in the delta of the Ganges and the Brahmaputra that is severely flooded every five or ten years. The vulnerability of such populations—often among the poorest in the world—to natural hazards is therefore rising because of human factors. The population density in coastal zones is about three times the global average, which explains the high human toll of the 2004 tsunami in Asia.

Natural Variability vs. Human Impact

The earth is far from being a stable, static system in equilibrium. Tectonic plates shift. The mountains rise by a few millimeters per year. Volcanic eruptions alter the landscape and modify the composition of the stratosphere by projecting great quantities of ash into it. Variations in the earth's orbit around the sun trigger glacial and interglacial cycles lasting some tens of thousands of years, with an associated variation in temperature of several degrees. The climate also varies on a millennial scale, with fluctuations of temperature that are more rapid but whose amplitude is smaller. In interglacial periods, such as the one in which we are living today, natural climatic variations on the scale of a century or a decade are common. Each region of the globe responds in a particular way to these variations: the so-called Little Ice Age that lowered temperatures in Europe between 1300 and 1850 is an example. Still shorter climatic cycles, such as the one associated with the El Niño/Southern Oscillation in the Pacific Ocean, periodically modify temperature and precipitation in regions as distant as North America and southern Africa. Solar activity varies as well, exhibiting several cycles, one of which has a periodicity of about eleven years.

Glaciers advance and recede with the rhythm of these climatic fluctuations, carving out deep valleys in the earth's surface. Soils are shaped and eroded by the action of rain, wind, and water flows. The eroded material is carried along over long distances by rivers and streams and deposited in floodplains. The landscape is therefore constantly being remodeled. Plant formation passes through various stages, the succession of which is regularly interrupted by natural disturbances: fires caused by lightning, invasions of pests, tornadoes, landslides, and so on.

Plant and animal species have come to be differentiated over millions of years, evolving by sexual recombination or genetic mutation in accordance with a process of natural selection based on their reproductive success. The rate of expansion of biological diversity has sharply varied in the course of the history of the living world. Five periods of massive extinction of biological species have been recorded, all as a result of natural causes (glaciations, asteroid or comet impacts, explosive volcanic eruptions). One massive wave of biological extinction 250 million years ago eliminated 90 percent of marine species, almost 70 percent of terrestrial vertebrates, and the majority of terrestrial plants over a period of 8,000 to 500,000 years.

The scientific challenge posed in recent decades by growing evidence of the human impact upon the environment has been to separate this impact from natural variability. Detecting the human signature hidden within a high level of natural background noise is now easier than it once was, for several reasons. First, technological advances have considerably improved both the quality and the quantity of data on the environment. Second, methodological advances (due in part to the analysis of core samples of ice obtained in polar zones) have yielded a better picture of the past evolution of the environment, particularly during the period prior to the one in which mankind began to exert a significant influence upon nature.

These new data have made it possible to test and refine models that seek to re-create the past state of the planetary environment. Additionally, human activity so profoundly transformed the terrestrial system during the twentieth century that many environmental parameters are today approaching values never previously attained. This departure from the historical range of the earth's natural behavior further facilitates the detection of human influence. The speed with which human activity modified the environment in the second half of the twentieth century is unique in the history of the planet. It cannot possibly be doubted that the natural world has been altered more rapidly during the last fifty years, and on a vaster scale, than during any other such period in the whole of human history.

Let us, then, briefly review the factual record of recent environmental changes. The details of this record are uncontroversial: they command a broad scientific consensus and imply no particular set of value judgments.

The End of Untamed Nature

Ever since mankind mastered fire and domesticated plants and animals, it has cut down forests in order to work the land. More than half of the earth's exposed surface (that is, land not covered by ice sheets) has been significantly modified in the course of the last ten thousand years: land in its "wild" state, untamed by human intervention, accounts today for only 46 percent of this exposed surface. In all parts of the world, agriculture has replaced natural vegetation in order to meet the growing needs of the world's population for food and fiber. Even those regions that still remain undisturbed suffer the indirect effects of human activity through the deposition of atmospheric pollutants emitted by industry and intensive agriculture. Fifteen percent of the world's wetlands and marshes have been drained. Between 10 and 50 million hectares (roughly 40,000 and 200,000 square miles) of land have been reclaimed from the sea as a result of coastal development. Human activity, agriculture in particular, is responsible for between 60 and 80 percent of soil erosion throughout the world.

The global extent of cultivated land rose from 350 million hectares (1.35 million square miles) in 1700 to 1.65 billion hectares (6.35 million square miles) in 1990—an almost fivefold increase over three centuries. During the same period, the amount of land devoted to livestock farming has increased by 500 million hectares (1.93 million square miles). The greatest concentration of farmland in the world is found in Eastern Europe, where more than half of the land is under cultivation. During the twentieth century, the cultivated area of the planet increased by 50 percent, principally in the tropical regions. Little land is now left for agricultural expansion in developing countries, where the demand for food is rising, either because uncultivated areas are covered with tropical forests or because their soils are poorly suited to agriculture. In Western Europe and in the northeastern part of the United States, by contrast, the total area under cultivation in the course of recent decades has shrunk with the abandonment of agriculture in marginally productive regions. Worldwide, 222 million hectares (857,000 square miles) of land have been given over to uses other than farming since 1900.

Whereas world food production virtually doubled between 1961 and 1996, the area under cultivation increased only by 10 percent during this period. By contrast, the extent of irrigated lands—a major factor in global water consumption—increased by 70 percent, reaching a total of 271 million hectares (1.05 million square miles) in 2000. Two-thirds of the world's irrigated agriculture is in Asia. More than 10 percent of the land used for this purpose suffers from salinization—an increase in the concentration of salts in the soil as a result of poor irrigation management, the effect of which is to reduce the land's productivity. On a global scale, the utilization of nitrogen- and phosphate-based fertilizers increased by a factor of 7 and 3.5, respectively, between 1961 and 1996. Intensive agriculture currently uses more nitrogenous fertilizers than the total amount of nitrogen naturally fixed by all of the earth's ecosystems. In 1990, 150 million tons of artificial chemical fertilizers and 3 million tons of chemical pesticides were applied worldwide. Almost half of chemical fertilizers end up in watercourses, lakes, and underground water.

Deforestation is defined as the conversion of forests to other forms of land use and the reduction of tree cover below a threshold of 10 percent of the surface area. The United Nations estimates that the clearing of natural forests proceeded at an average annual rate of 16.1 million hectares (slightly more than 62,000 square miles) during the 1990s, that is, a loss of 4.2 percent in ten years by comparison with 1990. This is equivalent to the disappearance every three years of forestland equal roughly in area to the whole of France. The natural forest sometimes spontaneously regenerates on cleared plots of land. Stands of trees have been planted on land previously uncovered by forests as well, notably on abandoned farmland in Western Europe and the northeastern part of the United States, partly offsetting the loss due to deforestation. The net decrease of forested land worldwide between 1990 and 2000 was 9.4 million hectares (more than 36,000 square miles) annually between 1990 and 2000.

This decrease is a cause for grave concern in the case of tropical rain forests, which are the world's richest sources of biodiversity and play an important role in the carbon cycle. In these forests, the net rate of deforestation averaged 0.43 percent per year between 1990 and 1997; in some "hot spots," the rate is ten times higher.

In 2003, some 3 billion people, or almost half of the world's population, lived in cities. In all parts of the world, the urban population has increased more rapidly than the rural population. The number of cities with more than 10 million residents—"megacities," having a population equal to or greater than the population of Belgium—has jumped from one in 1950 (New York) to nineteen in 2000, most of them in developing countries. Conurbations—areas of continuous and dense urban development—are concentrated along coasts and major rivers in India and the Far East, along the East Coast of the United States, and in Western Europe.

Built and paved surfaces presently cover only about 2 percent of the earth's exposed surface, but each year they occupy 1 to 2 million additional hectares (roughly 4,000 to 8,000 square miles) of high-potential agricultural areas and, moreover, have a considerable ecological footprint. Urban atmospheric pollution produces high concentrations of ozone in a circumambient radius of more than sixty miles, as a consequence of a chemical reaction caused by solar radiation (low-altitude ozone significantly retards the growth of vegetation). The residents of cities along the Baltic Sea, for example, depend on a system of forests, agriculture, lakes, and coastal waters that extend over a region almost one thousand times greater than the urban area proper. The city of Dakar in Senegal draws its water from a lake situated some 125 miles away.

The Threat to Water Sources and Marine Life

The consumption of fresh water in the world was forty times greater in 1990 than in 1700. Between 1900 and 1975, the quantity of water extracted from natural sources rose from 579 billion to 3.765 trillion cubic meters per year. Of this latter figure for 1995, 2.265 trillion cubic meters were consumed and the rest was returned to nature, often after having lost much of its quality. The quantity of water extracted in 2000 varied from 1,932 cubic meters per person in the United States to 675 cubic meters in France, 1,011 cubic meters for countries such as Egypt that depend on irrigated agriculture, and roughly 10 cubic meters for the countries of Central Africa.

A growing proportion (almost two-thirds according to certain estimates) of surface water flows is blocked by dams and diverted by dikes or canals. Between 1950 and 2000, the worldwide number of large dams (more than 50 feet high) went from 5,700 to 41,000, affecting 60 percent of the great river basins. The Aswan Dam in Egypt intercepts 98 percent of the silt carried along by the Nile and prevents the nutritive elements of the river from flowing into the Mediterranean, which has led to a considerable decline in the stocks of shrimp and sardines where the Nile Delta empties into the sea.

In the course of the twentieth century, the consumption of fresh water increased by a factor of nine, with the result that certain large rivers such as the Nile, the Yellow River, and the Colorado River no longer reach the sea during part of the year. In Arizona the extraction of water from underground aquifers for human consumption occurs at twice the rate that these sources are replenished; in the city of Dhaka, in Bangladesh, so much underground water is pumped that the level of the water table has fallen by 130 feet in certain places (new wells produce three times less water than wells drilled thirty years ago). Worldwide, between 60 and 75 percent of the fresh water available for all purposes is used for irrigated agriculture. In 2000, more than a billion people had no access to clean drinking water, and 2.4 billion were served by sanitation systems inferior to the standard enjoyed in ancient Rome. In developing countries in 1997, 90 percent of the water used flowed back into rivers without treatment. One hundred million city dwellers in poor countries habitually defecate outdoors, for want of toilets.

In 2000, 3 billion tons of fish were being taken from the world's oceans every year, or thirty-five times more than the total annual catch at the beginning of the twentieth century. More than a quarter of the biological production of the best-stocked regions of the ocean is now claimed by fishing. In 2000, 47 percent of the stocks of saltwater fish for which data were available were considered to be harvested to the maximum ecologically sustainable level, 18 percent were considered to be overharvested, and 8 percent depleted. The population of blue whales declined by 99.75 percent in the oceans of the Southern Hemisphere between 1890 and 1990. Other species of whales were relentlessly exterminated until 1986, when an international moratorium on commercial whaling was negotiated.

A systematic study of the ecological impact of commercial fishing, published in 2003 in the prestigious journal *Nature*, revealed that the oceans have lost more than 90 percent of the populations of large fish (cod, tuna, skate, swordfish, halibut, plaice, shark) by comparison with their preindustrial levels. Since the end of the 1980s, the quantity of fish removed from the oceans by the fishing industry has fallen by about 500,000 tons per year as a result of a decrease in the number of fish. Small fish, on which large fish feed, are now more and more frequently taken for human consumption.

A Voracious Appetite for Energy and Materials

In the nineteenth century, with the invention of the steam engine and the growing use of coal, worldwide energy consumption increased fivefold by comparison with the preceding century. This consumption grew still further in the twentieth century, by a factor of sixteen, with the exploitation of petroleum and natural gas deposits and later, less extensively, reliance upon nuclear energy. A rough calculation by the American environmental historian John McNeill indicates that the consumption of energy during the twentieth century as a whole was ten times greater than that of the previous millennium. Today about 95 percent of global energy use comes from fossil fuels: petroleum (44 percent), natural gas (26 percent), and coal (25 percent). Nuclear reactors account for 2.4 percent of world energy production. Renewable energy sources (solar, eolian, and geothermic, as well as biofuels) represent only 0.2 percent of world energy production, not counting hydroelectric dams (2.5 percent).

Between 1950 and 1990, world steel production quadrupled (reaching 773 million tons per year by 1990) and world paper production increased 5.5 times (270 million tons per year by 1990). By 1990, the American economy was directly consuming 6 billion tons of material a year, or an average of 110 pounds per person per day (a bit more than 500 pounds if the materials lost during the production process are factored in, as against 300 for the countries of the European Union and 270 for Japan).

The Rising Tide of Industrial Waste

In the course of extracting, transforming, and distributing materials and products, industrial activity generates more than 40 billion tons of solid waste per year. By comparison, the total quantity of sediments and other natural materials conveyed by the world's watercourses amounts to only 10 to 25 billion tons. The manufacture of a single computer semiconductor chip generates 630 times its weight in waste.

The average quantity of solid and liquid waste produced per capita annually is 0.3 tons in several European countries and 0.7 tons in the United States. A high proportion of this waste is considered to be toxic. Some types of nuclear waste will remain lethal for 24,000 years.

In 1992, plastics—which scarcely existed in 1950—represented 60 percent of the waste recovered from the world's beaches. With the advent of petroleum came oil slicks. Some have been due to accidents: in 1979, the *Atlantic Express* spilled 287,000 tons of oil in the Caribbean; the *Amoco Cadiz* had anticipated this record the year before, releasing 223,000 tons of crude oil off the coast of Brittany; in 1989, the *Exxon Valdez* polluted the coast of Alaska with 37,000 tons of crude oil; in 2002, the *Prestige* lost 17,000 tons of oil off the coast of Spain. The frequency of such large-scale disasters is diminishing, however, and most of the petroleum present in the world's oceans today is due to the flushing of oil tanks at sea by unscrupulous tanker captains.

Finally, hundreds of miles above the earth, space has become humanity's latest dumping ground. Several thousand disused satellites, rocket engines, and miscellaneous pieces of debris now orbit the planet at speeds of several thousand miles per hour.

In 2000, 160 million tons of sulfur dioxide (more than double the total of natural emissions) were emitted into the atmosphere by industrial activity—an increase of thirteen times by comparison with emissions at the beginning of the twentieth century. This pollution is responsible for the acidification of the rain and of lakes in highly industrialized regions. At the beginning of the twenty-first century, 6.5 billion tons of carbon (as opposed to 500 million tons in 1900) were expelled into the atmosphere. The atmospheric concentration of carbon dioxide (CO_2) exceeds the maximum values of the natural fluctuations in the concentration of this gas during the glacial-interglacial cycles of the last few hundred thousand years. The natural state of the atmosphere has therefore been altered. Twenty-five million tons of nitrogen are also emitted each year into the atmosphere in the form of NO_X; the atmospheric concentration of methane (CH_4) has increased by 150 percent since 1750 (a rise due mainly to the 1.4 billion head of cattle currently on the planet); and the concentration of nitrous oxide (N_20) in the atmosphere is higher than it has been for more than a thousand years.

Human activity releases numerous toxic substances into the natural environment, such as dioxins, whose rates of decay and long-term effects on health are poorly understood. Thus, for example, human activity was responsible in 1980 for 78,000 tons of arsenic escaping into the natural environment, four times more than the amount that is naturally released. Lead emissions have increased roughly eightfold during the twentieth century. Ice formed in the Arctic in the twentieth century contains almost 100 times more lead than the ice formed before this element began to be exploited (lead has been used in plumbing since the time of the Roman Empire). The concentrations of copper, zinc, mercury, and cadmium in nature have more than doubled since the pre-industrial era, with local soil concentrations reaching levels 10 to 100 times higher than before 1800. Emissions of non-toxic substances such as chlorofluorocarbons (freons and other halons) are responsible for the seasonal depletion of the ozone layer in the Antarctic and the Northern Hemisphere. One consequence of this has been an increase, during certain months of the year, in ultraviolet radiation on the earth's surface. High levels of this radiation affect the photosynthesis of green plants and kill phytoplankton in the oceans.

The First Signs of Climatic Warming

Greenhouse gases emitted into the atmosphere as a result of human activity have contributed to an increase in the temperature of the global atmosphere. Their effect in 2000 was equivalent to an increase of slightly more than 1 percent in the solar radiation reaching the surface of our planet. During the twentieth century, the average temperature on the earth's surface increased by about 2.1 degrees Fahrenheit. In the late twentieth century, according to the latest and most comprehensive study based on a range of past temperature records (or "proxies"), the Northern Hemisphere experienced the warmest and most widespread temperatures since 900 A.D. The first years of the twenty-first century have continued the trend, as the deadly consequences of the scorching summer of 2003 in Europe (particularly in France) attest. The ten hottest years in Europe out of the past five hundred years all occurred after 1989.

This climatic warming has brought with it a lengthening of the growing season and a shortening by about two weeks of the period during which lakes and watercourses in the boreal and temperate zones of the Northern Hemisphere are frozen. On a global scale, the area of the earth permanently or seasonally covered by snow has decreased by about 10 percent since the end of the 1960s. The Arctic ice cap has retreated by a million square kilometers (about 385,000 square miles) since 1978, a rate of 3 to 4 percent per decade. The thickness of the ice in the Atlantic Ocean has decreased by about 40 percent from late summer to early fall over recent decades.

The retreat of the glaciers can also be observed outside the polar zone. More than 80 percent of the ice accumulated at the summit of Mount Kilimanjaro, the highest peak in Africa, has melted since the beginning of the twentieth century. The average level of the oceans has risen by 4 to 8 inches during the twentieth century. The frequency of extreme climatic events (severe storms, large atmospheric depressions, droughts) has increased in several, though not all, regions of the planet.

The Most Rapid Wave of Extinction Ever Recorded

The rate of extinction of animal and plant species is estimated to be at least several hundred times greater today than the natural rate. Establishing the extinction of a species is difficult, for the process is gradual and, once a species is extinct, only fossils or scientific records permit its disappearance to be confirmed.

At least 811 species extinctions over the last five centuries can be conclusively demonstrated. Depending on the methods employed, it is estimated that between 3.5 and 12 percent of bird species are threatened with extinction worldwide. For particular regions, the rate is much higher. Of 240 known primate species, nineteen are in danger of disappearing in the next twenty years, for their populations have fallen beneath the minimum threshold necessary for the perpetuation of the species. Ninety-seven others are threatened to various degrees by the disappearance of their habitat, notably from deforestation in tropical regions. Coral reefs and mangroves are rapidly being diminished or degraded in several places on the planet.

If these tendencies are confirmed, the most rapid wave of extinction that the earth has ever known is under way, solely as a result of human activity. Many experts agree that this extinction (the sixth in the history of the living world) is still more rapid than the one that probably was caused, 65 million years ago, by a collision between the earth and a giant meteor, which led to the disappearance of the dinosaurs. Assessing the exact extent of the extinction presently taking place is not yet possible, if only because the total number of species living on the earth remains uncertain and only a small fraction of these species has so far been described.

In addition to the processes of natural selection, mankind has introduced artificial processes that have shaped the earth's animal and plant populations. Species that fulfill human needs (domestic animals, cereals, and eucaplypti, for example) or that have succeeded in adapting to the environmental changes wrought by mankind (rats, the virus responsible for the common cold, and the tuberculosis bacillus, for example) have proliferated. In the course of the twentieth century, the populations of pigs, cows, sheep, goats, and chickens have increased still more rapidly than the human population. By contrast, species that are useful for mankind but difficult to domesticate (bison and blue whales), as well as ones that have not succeeded in adapting to a landscape dominated by mankind (gorillas and Siberian tigers), are becoming extinct.

Unprecedented Biological Invasions

From one continent to another, whether by accident or design, the development of international transport and large-scale human migrations have displaced animal and plant species. In some cases, these species have rapidly colonized their new environments, expelling native species from their customary habitats and dramatically altering the ecological balance. In 1889, campaigning in Somalia, the Italian army imported cattle suffering from rinderpest, a highly contagious disease that was then unknown in Africa. The mobility of livestock in East Africa led to the greatest epizootic (animal epidemic) in recorded history. Millions of head of cattle died, as well as millions of buffalo, antelopes, giraffes, wildebeest, and other wild ruminants. The sizable decrease in environmental pressure exerted by local herbivores led to a corresponding increase in the tree cover of the savannas of East Africa as well as a profound destabilization of the pastoral economies of the eastern and southern parts of the continent, with inevitable consequences for human beings: almost two-thirds of the Masai, among other tribal peoples, perished.

In the twentieth century, biological invasions of this type occurred with unprecedented frequency. After their introduction in Australia in 1859, the number of rabbits rose to roughly 500 million in that continent by 1950, to the great dismay of local sheep breeders, who found their grazing lands invaded. Myxomatosis, a highly infectious viral disease deliberately introduced as a countermeasure, succeeded in eliminating 99.8 percent of the rabbit population. But as new generations acquired an immunity to the disease, the population subsequently rebounded, reaching some 100 million in the 1990s. Other biological invasions have affected plant species: in North America, for example, the chestnut tree was eradicated following the accidental introduction of a particular kind of fungus.

One of the fears arising from the use of genetically modified organisms in agriculture is that these plants—or the hybrids they are apt to form with wild plants—will acquire an advantage over their natural congeners and uncontrollably invade other habitats, particularly if genes introduced in the laboratory to confer resistance to herbicides propagate to the hybrid species.

Uncertainties and Regional Variations

For other forms of environmental change, however, scientific consensus is inevitably hard to come by, chiefly because reliable quantitative data on a global scale are lacking. This is true particularly of desertification (which we will consider in chapter 6) and the impact upon the environment of pesticides and the tens of thousands of other synthetic chemical products that are released into nature. The scope of the disturbance to phosphorus and sulfur cycles is not perfectly understood. For many of these environmental changes, local data suggest that the impacts on ecosystems and human societies are potentially significant, but a detailed global view remains out of reach for the moment.

Most of the data reviewed so far concern the planet as a whole and describe a general course of development over recent centuries. Because they mask variations in specific patterns of evolution over regions and periods, it needs to be kept in mind that any generalization on a worldwide scale for the twentieth century unavoidably conceals divergent and complex tendencies.

Since the 1980s, tropical deforestation has mainly been concentrated in fifteen or so "hot spots," where rates of deforestation higher than 2.5 percent per year are observed; in Europe, by contrast, forests have gained in area. Air pollution in the great cities of the developing world threatens the health of their inhabitants; in the cities of North America, Western Europe, and Japan, however, the air quality has significantly improved over the past thirty years, after three centuries of degradation. In particular, the atmospheric concentrations of sulfur dioxide (SO_2) and carbon monoxide (CO) have considerably diminished in these regions thanks to efforts at controlling automobile and industrial pollution. Detailed historical and geographical analysis is therefore indispensable if the nature of specific environmental problems is to be properly understood.

The Greatest Environmental Disaster of the Twentieth Century

The Aral Sea takes the prize for the greatest ecological catastrophe caused by human activity on a regional scale. Straddling the former Soviet republics of Kazakhstan and Uzbekistan, in 1960 it was the fourth largest body of fresh water in the world. Since then its area has diminished by more than 50 percent and its volume by 75 percent; its level has fallen more than 50 feet; and its salinity has increased by more than three times. In 1990, it became separated into two parts.

Beginning in the early 1960s, most of the water from the two rivers that feed the Aral Sea, the Amu Dar'ya and the Syr Dar'ya, was diverted from its natural course to irrigate 7 million hectares (about 27,000 square miles) of cotton (and, to a lesser degree, rice) fields, an area equal to that of Ireland. A canal almost 700 miles long running through the Kara-Kum Desert was constructed to bring the water to neighboring Turkmenistan. Before 1960, some 55 billion cubic meters of water flowed into the Aral Sea, a figure comparable to the volume generated by the Po in Italy; by the 1990s, this contribution was not more than a tenth of its historical value, when not reduced to a trickle during periods of drought. Between 30 and 45 percent of the water diverted for agriculture was wasted, by evaporation and seepage from irrigation canals as well as by over-irrigation, some fields receiving as much as 12,500 cubic meters of water when 3,500 cubic meters would have sufficed (even 2,500 cubic meters with a judicious selection of cotton species).

Shortly after its construction, breaches in the walls of the Kara-Kum Canal threatened to flood Ashkhabad, the capital of Turkmenistan. Now virtually deprived of sources, the Aral Sea was rapidly emptied by evaporation.

The ecological and human consequences of the drying up of the Aral Sea were many. The local climate changed, with higher temperatures in summer, weaker precipitation during the rainy season, and a shortening by two weeks of the growing season for cotton. The amount of snow that fell on neighboring mountains decreased, causing the amount that melted in the spring, feeding the Amu Dar'ya and Syr Dar'ya rivers, to decrease as well. In the 1950s, between 40,000 and 60,000 tons of fish were taken from the Aral Sea each year. Beginning in the 1970s, the fish population collapsed because of the growing salinity of the water, eventually shutting down the fishing industry. Tens of thousands of fishing-related jobs were lost. Twenty of twenty-four local species of fish became extinct. The salt deposits on the dried-out bottom of the sea were dispersed by the winds throughout the region, with the result that agricultural yields fell; grazing lands became barren within a radius of 125 miles around the Aral Sea; concrete and steel structures (electric pylons among them) were corroded; and the human population suffered from severe cases of ocular irritation.

The level of the aquifers adjacent to the Aral Sea fell by between 16 and 50 feet, and the water became salty. In the irrigated zones, by contrast, groundwater levels rose by several meters, causing salts to accumulate in the soil and turning agricultural land into desert. The vegetation of the alluvial forests, marshlands, and pastures was altered as a result. By 1990, almost half of the mammal species and three-quarters of the bird species present in the region in 1960 had disappeared. The people of the region, for their part, suffer from a variety of health problems, not only on account of the density of sand and salt in the air, but also because of the contamination of the soil and water by the large amounts of pesticide that had been sprayed on the cotton fields (as much as 48 pounds per acre) in order to combat the harmful insects that prosper in this hot climate. The concentration of DDT in the soil is, on average, between two and seven times greater than the maximum acceptable level, with local concentrations as much as forty-six times higher than this level. Lethal pollutants have found their way into the water supply as well. The authorities in Kazakhstan discourage breastfeeding, given the high proportion of pesticide residues found in mother's milk; the infant mortality rate in the region is now one of the highest in the world. Restoring the Aral Sea to its former condition may be an impossible task. For at least the foreseeable future, what was once a vast blue sea will remain a salt desert, abandoned by most forms of life.

This ecological disaster is the result chiefly of economically inappropriate development policies, but also of the arrogance of a political elite, aggravated by the complicity of engineers and scientists. In the 1950s, some local leaders declared that the emptying of the Aral Sea was the price to be paid for regional development and that over time cotton farming would compensate for the disappearance of coastal industries. The budgets allocated to the various government departments responsible for the project and to the region's collective farms were calculated on the basis not of actual levels of agricultural production, but of the scale of infrastructure needing to be built and of the investment in technology this would entail. Planners and builders therefore had no incentive to limit waste. The objective of the project—to make the Soviet Union self-sufficient in agriculture—was nonetheless achieved, with the USSR becoming the world's second largest exporter of cotton.

This is not an isolated case. The idea of building the Three Gorges Dam on the Chang Jiang (Yangtze River) in China sprang from the same desire to dominate nature by means of gigantic construction projects, despite their ecological and human cost, largely as a way of affirming the power of the nation and its leaders. The same is true for the Hoover Dam on the Colorado River in the United States, built during the Great Depression in the 1930s, and the Aswan Dam, constructed in Egypt under Gamal Abdel Nasser, who saw it as a symbol of Arab nationalism and as an enterprise worthy of the pyramids and his ancient ancestors, the pharaohs. All these great schemes had severe ecological consequences, even if they did not quite match the disaster of the Aral Sea.

An Unprecedented Worldwide Expansion

While the earth was undergoing these enormous ecological changes during the last one hundred years, how did humanity change? Here, pessimism gives way to optimism.

Human longevity and prosperity have continually increased in recent centuries. The average life expectancy of the world's population rose from twenty-four years in the year 1000 to sixty-six years in 2000. In the past three hundred years, this population increased in size by a factor of ten, exceeding 6 billion individuals by the beginning of the present century. Two thousand years ago, at the beginning of the Christian era, humanity numbered only about 200 to 300 million persons, comparable to the population of the United States today.

The world economy, measured by the gross national product of all countries taken together, was 120 times larger in 2000 than in 1500. Its unprecedented expansion is partly due to the growth of population during this period, but it is also the result of more productive technologies and more efficient methods of production. Adjusting for changes in purchasing power, average per capita income was nine times higher at the end of the twentieth century than in 1500.

According to a study by the Organisation for Economic Co-operation and Development (OECD), the world economy was more robust during the last fifty years than during any previous period. The average rate of economic growth was 3.9 percent between 1950 and 2000, versus 1.6 percent between 1820 and 1950, and 0.3 percent between 1500 and 1820. Total foreign direct investment increased by a factor of seven (in constant prices) between 1978 and 1998. Communications and trade in goods and services experienced considerable growth in the second half of the twentieth century, stimulating the diffusion of ideas and technologies throughout the world.

Between 1950 and 1990, the number of television sets rose from 45 million to 825 million and the number of radios from 226 million to almost 2 billion. Between 1993 and 1999, the number of persons connected to the Internet rose from 3 million to 200 million. The total number of motor vehicles (cars, trucks, and motorcycles) rose from less than a million in 1910 to 777 million in 1995. In 1990, 456 million tourists traveled abroad, whereas in 1950 only 28 million left their own country. In the course of the same forty years, the number of airline passengers rose from 23 million to more than a billion per year.

At the same time, gaps between the richest and the poorest countries on the planet also widened (though the disparity in income between individuals decreased in relative terms, largely as a result of rapid economic growth in China, ignoring disparities within countries). Perhaps the most shocking statistic is that the richest 20 percent produce and consume 85 percent of the total dollar value of goods and services, whereas the poorest 20 percent command only 1.3 percent of world economic production. The fifteen wealthiest persons on the planet are worth more than the gross national product of all the countries in sub-Saharan Africa. Even though the standard of living of the world's poorest people has increased in absolute terms, 1.2 billion people still lived on less than a dollar a day in 2000.

In Africa, where the rate of demographic growth now exceeds the rate of economic growth per capita, income has fallen since 1980 (after having risen from 1820 to 1980). This drop in income may also be due to the fact that the informal sector of the economy—the whole of unreported economic activities that escape national accounting—has grown to the point that it now encroaches on the formal sector.

Although the number of poor people on the planet remains unacceptably high, the general tendency has nonetheless been in the direction of a significant improvement in average standards of living and absolute levels of income for the world's poorest. Impressive progress has been made, and continues to be made, across a wide range of sectors of activity.

The price for this progress has been an exceptional transformation of the natural system. Human activity has more powerfully modified nature in the last century than natural forces have done over the past few hundreds of thousands of years, creating a situation without precedent in the history of the planet. Humanity has involuntarily embarked upon a vast experiment for which there is no central guidance, no long-term plan, no possibility of turning back, and no second chance.

The world is never the less beautiful though viewed through a chink or knot-hole.

HENRY DAVID THOREAU

2 Mankind and Its Environment

Throughout history, attempts to describe the interactions between human activity and its natural environment have always included two dimensions: the influence of natural forces on the development of societies (mankind's adaptation to nature) and the impact of humanity on nature (the transformation of nature by mankind).

These two dimensions are already found in the Old Testament. In the second chapter of Genesis, it says: "Then the Lord God formed man of dust from the ground, and breathed into his nostrils the breath of life; and man became a living being. . . . The Lord God took the man and put him in the garden of Eden to till it and keep it" (2:7, 15). In the first chapter, however, in another description of Creation, we read: "Then God said, 'Let us make man in our image, after our likeness'" (1:26); and after having created man and woman:

> God blessed them, and God said to them: "Be fruitful and multiply, and fill the earth and subdue it; and have dominion over the fish of the sea and over the birds of the air and over every living thing that moves upon the earth." And God said, "Behold, I have given you every plant yielding seed which is upon the face of all the earth, and every tree with seed in its fruit; you shall have them for food." (1:28–29)

In one passage, humanity has been brought forth from the earth and entrusted with the mission of protecting it; in another, humanity is placed above nature and permitted to exploit it in order to live.

This ambivalence is found at the heart of the present work as well: mankind profoundly alters nature to its profit while yet remaining fundamentally dependent on the services nature furnishes. It must therefore preserve the integrity of nature's basic functions.

In the dualistic perspective of the Judeo-Christian tradition, mankind's role as the master of nature rapidly gained the upper hand over its role as guardian and the natural elements came to be seen as having no other function than to serve humanity. Among Christian thinkers, this view was contested only by Francis of Assisi. The Greek, Roman, and ancient Oriental traditions, by contrast, all conferred a sacred value on natural elements. Similarly, almost all small societies, including ones in the West, have conceived of mankind as being a part of nature and of having the duty to preserve it in exchange for the right to take from it the goods necessary for its own survival. Their relationship was seen as one of peaceful commerce and mutual respect.

Ever since ancient times, scientific theories about the interactions between human activity and nature have both reflected and influenced popular attitudes toward the natural environment, shaping the way in which the natural world is regarded and, at the same time, providing a basis for ideologies seeking to regulate mankind's behavior toward nature. The modern scientific conception of these interactions, based on the observation that the natural environment places limits on the development of human societies, underwent a significant change in the twentieth century as recognition of mankind's growing capacity to modify its environment gradually gave way to acceptance of the view that certain modifications have potentially harmful consequences for nature and mankind alike.

Early Scientific Conceptions

Ancient societies commonly sought to explain the course of human affairs by invoking the motions of the stars. This primordial belief in a relation of cause and effect between the celestial environment and human behavior survives in the popular imagination today in the form of astrology.

The idea that climatic conditions determine the political organization of societies—a first approximation of modern scientific approaches—originated with the ancient Greek historians and philosophers, and later was developed by Montesquieu and other French thinkers of the Enlightenment. This line of thought received its fullest expression at the end of the nineteenth century in the doctrine of environmental determinism, which held that culture and human behavior were entirely the product of their natural environment: thus the English developed a great aptitude for navigation because they lived on an island; the Arabs were monotheists because they lived in vast empty deserts, which conjured up in their minds the idea of a single God; the Eskimos were a primitive nomadic people because of the harsh conditions of their Arctic habitat, and so on.

It rapidly became apparent, however, that the diversity of human behaviors could not be explained by the influence of the environment alone: the people of Tasmania, an island whose physical characteristics resemble those of the British Isles, never constructed boats; before their conversion to Islam, Arab tribes believed in the existence of a multitude of gods; and the aboriginal inhabitants of Alaska live today in a landscape dotted with sophisticated installations for the extraction of oil. Despite its lack of explanatory power, the theory of environmental determinism survived until the 1920s, later cropping up from time to time in the writings of a few uninformed authors.

A weaker version of this theory, sometimes called environmental possibilism, accepts that the environment is not directly responsible for the ways in which human societies evolve, but nonetheless insists that the presence or absence of particular environmental factors imposes constraints by permitting or prohibiting various paths of development. Thus the inhabitants of coastal regions have the chance to develop marine navigation, whereas peoples living in landlocked regions do not. Animal domestication can occur only in regions where mammals having certain domesticable characteristics are plentiful. Certain kinds of agricultural practices are adopted only where suitable climatic conditions, notably a sufficiently long growing season, prevail. These observations are hard to contest in the case of civilizations relying on rudimentary tools, but modern technology makes it possible to overcome certain natural constraints.

In *Guns, Germs, and Steel* (1997), the American physiologist and anthropologist Jared Diamond implicitly appeals to environmental possibilism to explain differential patterns of historical development of the regions of the world over the last thirteen thousand years. Why were the New World and Australia colonized by Europeans but Europe was not colonized by Incas and aborigines? The reasons, according to Diamond, have to do with differences from one continent to another in the abundance of animal species and domesticable plants, as well as geographic characteristics that favored human migration and the diffusion of technological innovations within and among continents. These factors determined not only the resistance of populations to pathogens, but also increases in food production and population, and the development of large sedentary, socially stratified, and technologically and militarily advanced societies. A half century earlier, in *A Study of History* (1947), the British historian Arnold Toynbee had developed a similar argument, namely, that certain societies had to confront greater challenges than others in coping with the natural environment, which encouraged their progress toward higher levels of civilization. Changes in the organization of human society are therefore to be interpreted as a succession of responses to particular environmental conditions.

The problem with environmental possibilism, even if it succeeds in showing that particular developments could not have occurred in this or that region of the world, is that it is incapable either of explaining why these developments appear (or fail to appear) where circumstances are favorable to them or of predicting when such occurrences will occur in the future.

Borrowings from Biology

Darwin's theory of the evolution of species exerted a decisive influence on the understanding of the interactions between nature and human activities. Ever since the publication of *On the Origin of Species* (1859), biologists have studied the mechanisms of adaptation in plant and animal species, seeking to explain the appearance of organs or particular behaviors on the basis of environmental characteristics. The diversity of terrestrial forms of life was shown to be the result of selective pressures exerted by the environment. Fifty years ago the American anthropologist Julian Steward relied upon an analogous concept, cultural ecology, in arguing that certain socioeconomic and cultural aspects of human societies are the consequence of adaptations to natural constraints, particularly with regard to technologies for exploiting natural resources, demography, and economic and social structure.

Unlike plants and animals, human beings have the capacity to consciously adapt to new circumstances. They define alternatives and choose among them on the basis of their knowledge and their ability to anticipate certain changes in their environment. In ancient societies, pragmatic adaptations of this sort were sometimes transmitted from generation to generation by means of religious taboos. The sacred character of cows insisted upon by Hinduism, for example, and the prohibition against killing them that followed from this were originally an expression of the value of cattle for the Indian people: the benefits associated with cows—whose male offspring was used as a draft animal and whose manure served as fertilizer and fuel—contributed significantly to the material well-being of the population. Protecting cows was therefore a cultural adaptation to particular ecological conditions.

This type of explanation applies mainly to societies living in relatively stable conditions whose impact on the environment is limited. In complex societies, the transformation of the environment by human activity requires that forms of social and political organization be constantly reconfigured to suit new environmental conditions. The rigidity of taboos and other religious and social codes of behavior thus became an obstacle to the adaptive capacity of societies. It is necessary, then, to consider the relation between society and nature in terms of dynamic interactions rather than as a matter of one-way influence.

In the 1960s, the concept of an "ecosystem" (defined as the set of living organisms and nonliving elements of the environment—soil, water, climate—that interact within a spatially defined area) was applied to the understanding of interactions between human activity and nature. The different components of an ecosystem are functionally related by flows of energy, matter, and information. An ecosystem has a complex structure, and its elements are characterized by great diversity. The human population is considered to be one among the many species, animal and plant alike, that interact within the ecosystem. The purpose of these interactions is to assure the survival of the ecosystem as a whole, not the survival of the human population at the expense of its natural surroundings.

This approach has been applied chiefly to the analysis of primitive societies, whose characteristics are treated as factors of a population's adaptation to the conditions of its ecosystem. Thus ritual practices involving sacrifice and the castration of certain young boys (among the Moriori in the Chatham Islands off the coast of New Zealand, for example), as well as tribal warfare in primitive societies, are seen as serving to maintain population density at a level compatible with the natural ecosystem. These adaptive strategies take into account fluctuations in the amount of food available from one year to the next as a consequence of unpredictable climatic variations.

In certain cases, factors internal to human societies interfere with the process of adaptation to environmental constraints. These include rigid institutions, a weak capacity for technological innovation, an oppressive political system, and precarious economic conditions. Unsuccessful adaptation reduces the chances of a society's survival and diminishes its ability to respond to extreme disruptions. Problems of "maladaptation," as the American anthropologist Roy Rappaport calls them, become particularly severe in complex societies, where the interests of an elite frequently diverge from those of the rest of the population.

Such maladaptations may explain the disastrous impact of a society upon its natural environment, leading sometimes to its own collapse. In some cases, where the mechanisms that once allowed the society to adjust to new environmental conditions no longer function effectively, the social factors that perpetuate existing economic and political institutions tend to persist. While this process works to the benefit of elites, it is apt to degrade the ecosystem and diminish the well-being of the majority of the population.

All the explanations described so far consider the interactions between human activity and the environment mainly in terms of human adaptation to natural constraints: nature influences human activity. Although they apply well enough to societies that have not acquired an advanced level of technological mastery, these explanations do not satisfactorily account for the profound transformation of nature by mankind. This transformation first began with the emergence of complex societies, several thousand years ago, and gained considerable impetus accelerated with the advent of industrialization in the late eighteenth century.

In the mid-nineteenth century, the American lawyer and conservationist George Perkins Marsh—a member of Congress, his nation's ambassador to Turkey and Italy, and an attentive observer of the changes that mankind had wrought upon the natural landscape—published a detailed scientific study of the implications of human activity for the environment, *Man and Nature* (1864), that enunciated what are now considered to be the fundamental principles of contemporary environmental science. Indeed, Marsh is widely regarded today as the founder of the scientific study of the transformations of the terrestrial environment brought about by human beings.

The Influence of Neoclassical Economic Theory

The scientific conception of the interactions between human activity and the natural environment owes as much to Adam Smith's doctrine of the "invisible hand" as to Darwin's theory of evolution. In fact, an important question raised by theories of human ecology concerns the level at which adaptation to the environment occurs: is the relevant social unit of adaptation society—that is, a culture or population considered as a whole—or the individual? While biologically based theories tend to analyze adaptations to the environment at the level of populations, others, which lay greater emphasis on the transformation of nature by mankind, locate decisions affecting adaptation to natural constraints at the level of individuals.

One influential modern theory of collective behavior, inspired by neoclassical economics, assumes that rational, perfectly informed individuals interested only in advancing their own interests make a series of decisions about the best way of interacting with the environment, choosing courses of action in response to the constraints it imposes and opportunities it offers them. These choices aim either at maximizing financial gain or utility under certain constraints or at minimizing the risks of famine given variable and unpredictable environmental conditions. The aggregate effect of these decisions, repeated a great many times, is to generate certain forms of adaptation to the environment and transformation of this environment.

The deforestation of vast territories, for example, results from a multitude of individual decisions to clear small plots of land. Each of these decisions depends upon a calculation that balances the costs implied by this action (the work associated with cutting down trees and cultivating the soil) against the benefits that flow from them (income from the sale of wood and agricultural production). Certain actors make economically rewarding choices (clearing plots of land in wooded valleys and planting soybeans, which fetch a high price in the market), whereas others fail to recover their investment (clearing hillside plots, where the soil is quickly eroded, and planting coffee, whose price is falling). The unsuccessful strategy is therefore not imitated.

Rational choice theory assumes that each person's attempt to satisfy his own individual needs leads to a mode of interaction that best serves the interests of society as a whole. In the long term, these interests include the protection of the essential goods and services provided by ecosystems. The invisible hand of the market is supposed to automatically bring about the long-term protection of natural resources having value for human activity.

Today it is widely recognized that the private good may diverge from the public good under certain circumstances (as we shall see in connection with what is called the "tragedy of the commons"); that the information available to individuals is neither perfect nor equal; that certain actors who enjoy monopolistic power manipulate the market to their advantage; and that the value of a great many goods and services furnished by nature is not reflected in their market price. Market imperfections may therefore be at the root of environmental degradation.

Moreover, individuals are not wholly free from external influence in formulating their choices. Individual choices are the expression of preferences that reflect values associated with a particular social and cultural context. This context transcends economic actors and preexists them. In other words, a social system offers individuals a limited menu of strategies for adapting to the environment. In cultural contexts that promote unrestrained consumption, individuals will be able to drastically reduce their consumption only at the price of social marginalization. They risk therefore perpetuating an abusive relationship with the natural environment, even if this is contrary to their personal values.

Nonetheless, the explanation of environmental changes in terms of the decisions of individual actors and their interactions remains an important step forward from the theoretical point of view: the individual does in fact represent the elementary unit of decision making and interaction with natural ecosystems. In certain cases, those individual choices that lead to the most efficient forms of adaptation to the environment are progressively institutionalized in the form of cultural norms.

Politics and the Environment

Political ecology—the study of political struggles for control over natural resources or of political struggles whose outcome is determined by differential access to natural resources—explains ecological changes in terms of constantly evolving interactions (which is to say, a dialectic) between society and natural resources, as well as between social classes and various groups of resource users. From this perspective, environmental changes are the result of a set of social relations between, on the one hand, groups of managers of natural resources who have divergent interests and, on the other hand, these managers and larger political and economic entities, among them the market, the institutions that control access to resources in a given society, systems for the extraction and distribution of wealth, and the state.

The degradation of natural resources is therefore explained by shifting alliances between the state and private interests, which, via subsidy and resource allocation policies, reinforce the predatory exploitation of natural resources and widen the gap between elites and marginal segments of the population. Deforestation, for example, is seen as arising from the granting of privileged access to the forests of a country to logging companies, with the complicity of a corrupt government. These companies, often foreign, plunder the natural resources of a country, sometimes degrading them over the long term, without local populations benefiting in any significant way; indeed, indigenous peoples are liable to be evicted from the most richly endowed woodlands.

On this view, then, the interactions between human activity and the natural environment are largely influenced by social relations, institutions that regulate access to natural resources, and ideologies that orient the exploitation of these resources within the limits of what the physical environment allows. The course of socioeconomic evolution determines which productive natural resources, at a given time and place, will be the object of attempts at appropriation by different (and often rival) social groups. The theory of political ecology has some of its roots in the Marxist tradition, to the extent that it considers the exploitation of natural resources to be a function of social, economic, and political processes operating both within and across individual societies around the world. Certain contemporary movements that stress the globalization of economic activity in relation to the environment acknowledge the influence of this theory.

Applying Systems Theory to the Environment

Whereas the natural environment was seen at the beginning of the twentieth century as an essentially fixed collection of elements imposing limits on human activity, it is now conceived as a complex and dynamic system with which human societies maintain relations that are themselves constantly evolving. I find a number of the concepts of systems theory particularly useful in trying to understand human-environment interactions and therefore discuss them in some detail below. It should be emphasized that none of the theories I have just discussed above is inconsistent with a systems approach; indeed, linking up these various theories with one another is apt to be particularly fruitful.

General systems theory, derived from cybernetics, defines a system as a complex network formed of multiple elements, each of which fulfills specific functions and interacts with every other element. In a system the cause of a given change cannot be assigned to a particular element, for the change results from multiple interactions among all the elements of the system. The earth and humanity jointly form a vast and complex system, made up of a great many hierarchically organized subsystems. An organism, a human being, a forest, a business enterprise, a country—all these things are systems.

Terrestrial vegetation, the atmosphere, oceans, glaciers, geophysical processes, and human activity all interact with one another under the influence of a series of biogeochemical cycles that integrate them in a system of closely interdependent components. Water, for example, is stored in great quantities in glaciers and oceans, and evaporates from the oceans into the atmosphere, where it is carried over long distances and falls back to earth in the form of precipitation. Water is indispensable to animal and plant life, and circulates over the earth's surface in rivers and streams that displace great quantities of sediments. It is also used for agricultural and industrial purposes (notably irrigation), in addition to being directly consumed by human beings.

Carbon also has a cycle that involves the whole of the terrestrial system, passing from the atmosphere, where it contributes to the greenhouse effect, to the calcareous deposits of the oceans, and thence into vegetation, where it serves as the principal construction material for new growth, and the soil, where it is found in large quantities. Human activity significantly intervenes in the carbon cycle through the extraction of coal and oil reserves, which, in the absence of industrial activity, would remain isolated from currently active biogeochemical cycles.

The burning of these fossil energies releases carbon dioxide into the atmosphere, reinforcing the greenhouse effect and causing a global climatic change. This change particularly affects biological activity on land and in the oceans, and therefore the fixation of carbon by marine organisms and vegetation. Human activity also modifies the stocks and flows of carbon associated with vegetation by burning forests, planting trees, and altering the composition of plant species present in the landscape. International agreements such as the Kyoto Protocol to the United Nations Framework Convention on Climate Change, adopted in December 1997 and implemented since 2005, attempt to limit emissions of carbon dioxide and thereby control the human component of the carbon cycle.

Industrial and agricultural activities, along with a variety of political and economic institutions, influence the functioning of the earth today in much the same way as physical, biological, and chemical processes. In an age of increasing globalization in which the international socioeconomic system strongly interacts with the natural planetary system, the complexity of human systems has grown considerably, without, however, having yet attained quite the same level of complexity as nature. Nonetheless, just as cycles of water, carbon, and nitrogen traverse all the compartments of the terrestrial system, the circulation of information, merchandise, and people through worldwide socioeconomic networks is accelerating.

One of the most important characteristics of systems is their nonlinear behavior. Changing the contribution of a particular factor by a constant amount does not necessarily lead to a fixed change in the response of the system: once certain thresholds are exceeded, the state change of the system is speeded up or slowed down.

In Africa, an increase in population is accompanied by an expansion of small-scale subsistence agriculture and deforestation. In certain regions of the tropical forest in South America, the creation of large commercial farms at the expense of forestland is followed by a decline in the local population, which is forced to leave. In some European countries, the population declines and marginal farmland is permanently abandoned, whereas in parts of North America the abandonment of farmland is associated with an increase in population at the national level. What accounts for these apparently chaotic tendencies? Simply the fact that the size of the population interacts with many other factors—production technologies, the type of crops under cultivation, market demand, agricultural policies, cultural preferences, the supply of available labor, and so on—to determine the amount of land devoted to agriculture. These relations are systemic: production technologies depend on market demand, which is itself a function of population, which determines the number of available agricultural workers, whose level of education influences production technologies, and so on.

This sort of complexity, inherent to all systems, generates nonlinear responses to disturbances, which makes it difficult to predict outcomes. Surprises are possible. In particular, change is apt to accelerate suddenly once a certain threshold value has been exceeded. Insult your neighbor three times and it may be that, out of politeness, he will not react; insult him a fourth time, however, and the reaction will be disproportionate to the offense given, for you will have pushed him beyond the threshold that assures a certain measure of restraint. After a certain point, the response is not strictly proportional to the external stimulus. The terrestrial environment reacts similarly to the repeated insults inflicted upon it by human activity.

The complex behavior of systems arises in part from the existence of feedback mechanisms. Such mechanisms come into play when the state of the system affects its rate of change, which is thus to some degree independent of external influence. A positive feedback accelerates the rate of change. Thus, for example, a snowball rolling down a slope increases in size at an ever-faster rate: the greater its volume, the more additional snow it accumulates with each rotation. A negative feedback, by contrast, damps the rate of change and reinforces the system's stability: the flushing system of toilets includes a valve mechanism that slows the rate of water flow as the tank fills up, then cuts off the flow once the original water level is restored.

Positive and negative feedbacks often act upon a system simultaneously. Thus, for example, a street artist goes unnoticed until a few passersby stop to see what she is painting, at once attracting a growing number of curious onlookers (positive feedback). But as soon as spectators at the back of the crowd are unable to see the artist at work, they cease to linger and continue on their way (negative feedback). Through the establishment of an equilibrium between these positive and negative feedbacks, the population of onlookers is maintained at a stable level even though new passersby are continually stopping to take a look while others are leaving. The size of this stable population represents the threshold level (or carrying capacity) of spectators, given the particular configuration of the site in question. In the same way, the most efficient natural and social systems are those in which a robust equilibrium exists between several internal forces, some of which promote change and renewal while others preserve stability in the face of external shocks.

In a population where the number of births is higher than the number of deaths, demographic growth accelerates exponentially, for a larger population gives rise to a greater number of births (even if the number of children per couple remains constant) and a greater number of births increases the size of the population (positive feedback). With a rate of growth of 1.5 percent per year, a population doubles in forty-five years. If the initial population is 6 billion persons, 80 million additional members are added to the population in the first year, while in the forty-fifth year this figure is 160 million.

Negative feedbacks can moderate the evolution of the system beyond a critical threshold for population in relation to the environment. When an animal population becomes too large, the resulting scarcity of resources increases the level of mortality. In the history of human societies, this negative feedback has chiefly involved the control of births, modification of the rules authorizing marriage, emigration to other territories, and redistribution of resources within the population. It should be noted that a positive feedback may also suddenly accelerate a process of extinction—for example, when the population of an endangered animal species becomes so reduced that it is difficult for an individual to find a mate with whom to reproduce. Positive feedbacks give rise to vicious circles, while negative feedbacks promote stability.

Positive and negative feedbacks are at work in all natural and human systems, generating natural, economic, and sociopolitical cycles. The Canadian ecologist C. S. Holling has shown that the majority of these cycles are organized in four main phases: a phase of growth, which leads to an expansion and an increase in complexity under the influence of positive feedbacks; a phase of equilibrium, or stability, dominated by negative feedbacks, which introduce a certain rigidity in the system; a phase of disintegration, when the system collapses following an external shock that triggers positive feedbacks, which in turn push the system beyond its range of stability; and a phase of reorganization, during which a new structure is established that draws the system back into one of many possible stable states. Chance may determine the direction taken during the course of this phase, and therefore the trajectory followed during the whole of the subsequent cycle.

The climate is characterized by numerous cycles, of which the best known is the El Niño, with a periodicity of seven to eight years. The modern world economy undergoes Kondratieff cycles, whose longest periodicity is on the order of fifty years. Natural ecosystems are governed by cycles of plant succession, interrupted episodically by external disturbances such as fires, climatic catastrophes, and invasion by exotic species. Empires undergo decadence and collapse after a glorious period of expansion and apogee, with great variations in the length of the cycle. These cyclical changes are the result of a combination of processes internal to the system and external, abrupt, and episodic shocks.

When a society passes through a phase of growth or equilibrium, both its people and their leaders repeatedly commit the error of supposing that the current situation will go on indefinitely. This conviction creates a great reluctance to adapt their behavior to new circumstances or to anticipate a probable crisis. This lack of flexibility—lack of a will to change—only serves to precipitate the crisis. The end of the period either of growth or stability inevitably brings with it surprise and disappointment.

The timely implementation of adaptive strategies, on the other hand, is likely to diminish the severity of the crisis, minimize the pain that accompanies a rupture with the old régime, and soften the transition to a new form of organization, which then comes about in a voluntary and controlled way. This ability to anticipate future events is a tremendous asset, reserved to human beings, who enjoy a capacity for analysis and inference that is denied to climatic systems and ecosystems. It is nonetheless incumbent upon human societies to draw the fullest possible benefit from their intellectual endowments, something they are not always able to do in the face of forces acting in defense of hereditary privileges and acquired advantages, which introduce a degree of rigidity in the system that sometimes proves to be fatal.

Instability and Coevolution of Natural and Human Systems

Even if human societies and natural ecosystems are closely related, they form two quite distinct subsystems within the larger terrestrial system. Each is animated by its own dynamic and governed by specific rules. In both cases, change is the norm. Even before the appearance of life on earth, the planet was far from static: geological processes sculpted the earth's surface, wind and water eroded the soil, glaciers carved out valleys. Biological evolution, which gave rise to an extraordinary diversity of plant and animal life, and later the social and cultural evolution of humanity, carried this saga of perpetual motion forward. The familiar notion of the terrestrial system as an example of static equilibrium is pure fantasy: change is essential to the survival of the majority of animate complex systems. A river goes on being a river because upstream it is constantly fed by brooks and streams, and because downstream its waters are absorbed by the oceans.

Natural and human systems coevolve, which is to say they change together: a change in the behavior of one system introduces a modification in the environment of the other, forcing it to change as well. Nature and human societies are complex adaptive systems, frequently pushed beyond equilibrium in search of new forms of stability.

Until the nineteenth century, the natural environment's role in influencing the social and economic organization of human societies was predominant. In the twentieth century, technology gained the upper hand, and by its growing capacity to profoundly transform daily life and the environment came to have still greater impact. The growing risk of global environmental crises increases the likelihood that the present century will see the balance redressed yet again, with the environment regaining its former advantage. Political, economic, and social institutions may nonetheless be expected to play a decisive role in implementing novel strategies devised to meet a renewed round of environmental challenges created by human activity.

The high incidence of malaria, a tropical disease responsible for between 1 and 3 million deaths per year, well illustrates the phenomenon of coevolution, in this case between mosquitoes and the inhabitants of tropical regions. The characteristic parasites of malaria (*Plasmodium falciparum*, *vivax*, *malariae*, and *ovale*) are transmitted to human beings by certain mosquitoes, among them the notorious *Anopheles gambiae*, which thrives in Africa. In the mountainous areas of Vietnam, the indigenous tribal populations have never suffered to any great degree from malarial disease; indeed, they have adapted their housing in such a way as to limit the risks of being bitten by the mosquitoes that carry the parasite. These mosquitoes attack mainly at night, fly near to the ground, prefer cattle to humans, and flee smoke. The traditional homes of the mountain people are raised above the ground, with cattle being placed underneath the houses during the night and cooking done inside in order to produce smoke.

The adoption of such an efficient strategy for combating malaria is particularly surprising in view of the fact that the mountain people of Vietnam had not established a connection between mosquito bites and the disease. On the contrary, until recently they were convinced that malaria was caused by evil spirits or contaminated water. A series of adaptations over the course of centuries led them to notice that families who had organized their domestic arrangements in this fashion had a greater chance of surviving than others. Cultural tradition in this case was the product of human adaptation to nature, arising not from conscious, intentional decisions but from a series of small, more-or-less random cultural changes, followed by the replication of the most successful strategies.

Later, however, when the mountainous areas of Vietnam were colonized by ethnic groups from the lowlands, who laid out their homes in a different manner and who did not benefit from several centuries of cultural adaptation to this environment, the incidence of malaria rose sharply.

In 1939, scientists invented dichlorodiphenyltrichloroethane (DDT), an effective and easily produced insecticide against mosquitoes carrying malarial parasites. National campaigns, supported after World War II by the World Health Organization, promoted the spraying of DDT on the walls of houses, where the mosquitoes typically congregated. One application of DDT every few months proved to be sufficient to considerably reduce the mosquito population. These campaigns began in the 1950s, with the result that malaria had almost disappeared by the end of the 1960s.

A few years later, however, the mosquitoes reappeared, and malaria with them, infecting some 400 million people every year. This resurgence is explained by the mosquitoes' adaptation to the new environment created by human intervention. On the one hand, some mosquitoes benefited from a genetic mutation that rendered them impervious to DDT and so gave them a considerable reproductive advantage, by comparison with their congeners, in a new environment rich in DDT; and since, in most species of mosquito, several generations succeed one another in the course of a single season, this DDT-resistant gene was rapidly transmitted throughout the mosquito population. On the other hand, mosquitoes carrying malarial parasites adapted their behavior as well, no longer choosing the walls of houses as a resting place but selecting instead the surrounding vegetation, where DDT could not be sprayed (the discovery of the harmful effects of DDT on bird populations and the persistence of derivative chemical substances in the food chain finally led to its being banned in many countries during the 1970s).

In the next phase of coevolution, mankind once again modified its strategy for fighting malaria in order to respond to the genetic and behavioral evolution of mosquitoes. Since malaria affects human beings through the saliva of mosquitoes, which are simply vectors of the parasite (that is, its means of transport from one host to another), antimalarial medicines (such as Chloroquine) were developed to attack the parasite directly. Here again, the parasite found a way of parrying human initiative by coevolving through genetic mutations. Although resistance to Chloroquine did not appear until twenty years after its use on a large scale, only a few years were needed for the parasite to develop a resistance the next generation of medicines—a pattern of response that continues still today with the advent of new methods for fighting malaria.

Simpler strategies have been used successfully for decades. Thus, for example, families at risk of contracting the disease remove, to the fullest extent possible, all pools of water in which the mosquitoes reproduce—muddy puddles in the ruts left by truck tires, tanks filled with drinking water in homes, and so on—and sleep at night under insecticide-treated bed nets. More recently human societies have equipped themselves with a new and powerful tool in their adaptive struggle against malaria: the deciphered genome of the mosquitoes carrying the disease and of the parasites responsible for it. Biotechnology laboratories are working to develop a genetically modified mosquito that would be incapable of passing along the disease to humans, thereby interrupting the cycle of transmission.

Emergent Properties of Complex Systems

Another remarkable property of complex adaptive systems such as nature and human societies is that a multitude of small adaptations on the part of the components of these systems can give rise to a reorganization of the system as a whole, whose behavior then exhibits what philosophers call emergent properties. Consciousness, for example, is an emergent property of human physiology. Similarly, a change in the natural environment of the planet is an emergent property of local interactions between human activity and the physical, chemical, and biological processes that regulate nature.

While individuals are accustomed to reckoning the consequences of their actions with reference to the hierarchical level they occupy in a system (one's family, business, or immediate natural environment), it is more difficult to anticipate the consequences that will follow from a large number of similar actions at higher levels (the population of a country, the world economy, the global environment). In certain cases, the sum of a large number of individual actions may lead, unconsciously and unintentionally, to an unexpected and undesirable effect (demographic explosion, economic crisis, tropical deforestation, health-threatening atmospheric pollution, and so on).

This phenomenon of emergence at higher organizational levels of the terrestrial system is one of the reasons why a great effort must be made to understand the nature of environmental changes: the consequences of these changes need to be anticipated at the regional or planetary level. In the absence of careful scientific inquiry, an almost imperceptible reorganization of the complex system formed by nature and humanity could push it beyond a critical threshold, with the result that positive feedbacks cause a sudden acceleration of changes having potentially adverse consequences for the habitability of the planet by human beings.

An unpleasant surprise of this sort, though it represents a worst-case scenario, cannot be ruled out. Rapid reorganizations of the terrestrial system have already occurred in the past, when the planet had not yet been subjected to significant human disturbances. One such reorganization occurred twelve thousand years ago, with the retreat of the glaciers at the end of the last ice age during the climatic oscillation known as the Younger Dryas, which led to the abrupt cooling—in the space of one or two decades—of the climate of the entire Northern Hemisphere. The probable cause of this state change of the climatic system was a massive release of water from the melting of the glaciers. On reaching the ocean at the latitude on which Newfoundland lies today, this mass of fresh water slowed the Gulf Stream (the great warm current of the northern Atlantic) by diminishing its salinity, and therefore its density as well.

The Birth of a New Concept: Sustainable Development

Chris Patten, the former European Union commissioner for external relations, once described sustainable development as a matter of living on planet Earth with the intention of remaining there forever, rather than simply for a weekend holiday. Somewhat more formally, sustainable development may be defined as growth that satisfies our present aspirations without impairing the ability of future generations to prosper in accordance with theirs.

This concept was officially introduced in 1987 by the World Commission on the Environment and Development of the United Nations, under the direction of Gro Harlem Brundtland, a former prime minister of her native Norway and later director general of the World Health Organization. Since then, it has been widely adopted in civil society as well as in the spheres of business and politics, becoming a touchstone for public-interest organizations and institutions dedicated to promoting economic development without degrading the environment.

The very ambiguity of the expression "sustainable development" and the absence of any simple procedure for achieving its objectives are no doubt a large part of the reason for its popularity. Industrialized countries interpret it as requiring them to do a better job of managing the environment within their own borders, without compromising economic growth, and to work to slow the disappearance of tropical forests and endangered species beyond their borders. Developing countries, on the other hand, regard it as allowing them to assign priority to economic objectives rather than to environmental protection; in their view, eradicating poverty and catching up with the industrialized countries are preconditions for more responsible stewardship of the environment. A notable shift of consensus in favor of the latter position took place over the course of the decade separating the 1992 United Nations Conference on the Environment and Development (UNCED) in Rio de Janeiro, where priority had been given to the objective of protecting the environment, and the 2002 World Summit on Sustainable Development

(WSSD) in Johannesburg, where reducing poverty was put at the top of the global agenda.

Sustainable development is above all a matter of political choice, which inevitably means having to strike a balance between environmental, economic, and social objectives. It is not possible to devise an operational method for achieving sustainable development without specifying an order of preference among factors such as biological productivity, genetic diversity, climatic stability, the resilience of ecosystems (that is, their capacity to recover from shocks), the maintenance of a constant stock of natural resources, the satisfaction of basic human needs, growth in the supply of goods and services, cultural diversity, institutional stability, social justice, equitable treatment of different groups in society, and citizen participation in political decisions.

Any ranking of these factors is necessarily subjective and depends upon a particular social and cultural context. Consider the choice facing the automobile industry, where improving the safety of new cars often leads to an increase in their weight and therefore in their gas consumption and carbon dioxide emissions as well. Which objective is to be put first? And to whom does it fall to make this choice—the industry, consumers, the state, or future generations (represented by certain segments of civil society)? The problem is made all the more delicate by the fact that these various parties will profit to different degrees depending on whether improving safety is preferred to reducing pollution, or vice versa.

The concept of sustainable development is an attempt to reconcile two of the great myths of Western civilization: the promise of continual and indefinite progress, and the vision of a society living in harmony with unspoiled nature. By enjoining humanity to exploit nature, for the purpose of generating economic growth, and at the same time to act as its guardian, with the duty of maintaining nature in its original state, it is supposed that the contradiction found in Genesis concerning mankind's relation to nature can somehow be removed. But this merely restates the contradiction, without doing anything to eliminate it. In its simplest form, at least, the concept is both uninformative from the theoretical point of view, to the extent that it fails to provide a coherent account of the ways in which humanity and nature interact, and unhelpful from the practical point of view, being silent as to the means by which a noncontradictory regime of sustainable development could be instituted.

In spite of these defects, it has attracted considerable support. Part of the reason for this is that the concept of sustainable development is compatible not only with the logic of modernity, but with the whole rationalist and anthropocentric style of thought that has dominated industrial development for two centuries, while pushing this logic beyond its usual boundaries. It rejects the idea of zero economic growth and insists instead on the necessity of formulating an appropriate role for market forces in economic development, reforming the political process to take into account the interests of individual citizens, and eradicating poverty, often considered to be one of the major causes of the degradation of natural resources. Nothing in any of this is likely to frighten members of the Establishment. The pragmatic appeal of the concept of sustainable development arises from the fact that it holds out the prospect of being able to manage growth more effectively, not by arresting economic development, but by rationalizing it in such a way as to limit the harm caused to the environment and society.

In the last twenty years, economic science and ecology have collaborated in a joint attempt to flesh out the concept of sustainable development. Economists have borrowed from ecology, creating a new field of study within the neoclassical tradition known as environmental economics; ecologists, for their part, have taken an interest in economics, developing a new field called ecological economics that goes beyond the narrow postulates of mainstream economic theory. One of the fundamental aims of these two disciplines is to estimate the financial value of the services that nature provides to mankind and to identify market mechanisms by which this value can be taken into account in economic policy making. Current research is devoted to creating economic incentives for incorporating the value, or real importance, of the services that ecosystems provide in the prices of goods. While the services supplied by the environment are often considered to be public—and therefore free—goods, a compelling argument can be made that the users of these services should pay for them, particularly when this use diminishes or degrades a given resource, and that those who bear the financial burden of protecting and maintaining natural resources should be adequately compensated.

Federal or state regulations could be imagined, for example, that would require factory owners to compensate neighboring populations for damages suffered from pollution, encouraging owners to control emissions in order to keep costs down. A public institution might agree to pay the owners of land in a wooded catchment area—as New York City has done for more than ten years now—in exchange for a promise not to convert their property to other uses that would have the effect of reducing the role of the natural ecosystem in filtering water.

Establishing a market that integrates the environmental impact of production and consumption would require economic policies based on taxes and subventions (commonly called ecotaxes and ecobonuses), legal regimes that assign ownership of ecosystems and their services to defined entities, the commercial trading of pollution permits and rights to use scarce resources (tradable fishing quotas, marketable hunting permits), and the treatment of environmental degradation as a liability (in the accounting sense) in assessing the financial performance of a business or a country. In the phrase of the American biologist Edward O. Wilson, a green thumb needs to be added to Adam Smith's invisible hand.

For more than a decade now, political and business leaders have urged companies to embrace the imperatives of sustainable development, emphasizing the social responsibility of private enterprise (voluntary action by firms to integrate social and environmental criteria in their operations), the need to adopt a strategy of ecoefficiency (improving the quality of the environment while at the same time increasing profits by reducing costs), and the "triple bottom line" (an overall measure of a firm's performance in its economic, environmental, and social aspects). This has given rise to concrete measures such as waste recycling, investment in research on "clean" technologies, the ecological certification of sensitive resources, and the marketing of organic products.

The success of these innovations will depend on achieving a judicious balance between new government regulations and measures voluntarily adopted by businesses, as well as between sanctions and incentives (so that institutions endorsing sound ecological and social practices may expect to find their image enhanced in the public mind). An overly inflexible and constraining regulatory context would be counterproductive, for it would only hasten the relocation of enterprises to countries with more relaxed environmental laws. Conversely, it would be naive to suppose that all businesses will voluntarily accept restrictions aimed at promoting cleaner and more equitable forms of economic development.

On balance, then, the concept of sustainable development poses more questions than it answers. Is an environmental crisis the reflection of a deeper crisis of social organization and of the underlying cultural values of a society? Or is it simply a problem that can be overcome by inventing more effective technologies? If the latter, the solution would appear to lie in quickening the pace of technological innovation and adapting forms of social organization accordingly. Or is it necessary, by contrast, to tame the insatiable appetite for greater wealth and power, by profoundly modifying the cultural values that stimulate it?

Examining a few simple concepts and models of the causes of environmental degradation will help us think about these questions more intelligently.

> . . . to carry on my reflections in due order, beginning with objects that were the most simple and easy to understand, in order to rise little by little, or by degrees, to knowledge of the most complex. . . .
>
> **RENÉ DESCARTES**

3 The Mechanisms of Environmental Degradation

Reviewing the various scientific conceptions of the interaction between human activity and the natural environment is important because it gives us a better sense of the profound nature of contemporary environmental change. Beyond these general conceptions, however, we need to know which mechanisms are responsible for the alteration of particular natural environments. What are the most important causes of environmental change? How do they interact to produce it?

The analysis of particular situations leaves one with the impression that the causes of environmental change are extremely complex—so complex, in fact, that inferring general laws from the apparent confusion of local events seems a daunting task. Considering a few general models of environmental change will help us to detect certain regularities that obtain across a wide range of individual cases. It needs to be kept in mind that a model is a simplified and idealized representation of the world. It does not pretend to capture the whole of reality. A model abstracts certain features that are of particular interest, in this case the most important mechanisms of environmental change.

Historically, the development of these models followed a logical progression: elementary models were constructed with two or three variables that interacted in simple ways, and then modified to incorporate a greater number of variables that interacted in more complex ways. None of these models claims to represent the entire set of processes that lead to a particular environmental change. Each of them, however, brings out an important aspect of the relationship between human activity and its natural environment.

First Model: The Concept of Capital

Let us begin by formulating the requirements of sustainable development on the basis of the concept of capital, defined as a stock of wealth that generates a flow of goods and services. This wealth assumes several forms. The first, artificial (or constructed) capital, is made up of machines and infrastructure produced by human activity: cars, factories, houses, highways, and so on. The second, social capital, is constituted by the set of institutions, interpersonal relations, rules, and norms that make life in a society possible: legal, educational, and political systems. The third, human capital, consists in each person's capacity for work as well as the body of knowledge accumulated by human societies: culture, science, and other forms of learning.

The survival and development of humanity also depend on a fourth category of wealth: natural capital. The ozone layer filters ultraviolet rays, which otherwise would have a disastrous effect on health. Forests regulate the water cycle and provide a habitat for animal and plant species that are potentially useful for the development of new pharmaceutical products, while the nutritive elements of the soil make agricultural production possible. Natural capital also includes mineral deposits and other natural resources that help economies function.

Certain goods and services furnished by nature, such as petroleum, are the object of commercial exchange; others, such as air, are not. Poor countries depend more directly on their natural capital than industrialized ones. Peasants in the Sahel, in Africa, take their water directly from a nearby river, whereas in industrialized countries water is conveyed by a sophisticated system of purification, collection, filtration, and distribution. Until recently, artificial and human capital were the principal factors limiting the economic development of human societies. Today natural capital is a limiting factor as well.

As a theoretical matter, supposing that these different forms of capital could be quantified in terms of a common unit, three types of sustainable development are possible. Let us consider them in turn.

Perfect Substitutability of Capital

The first type assumes that natural and artificial capital are perfectly substitutable for each other. On this assumption, it is permissible to clear the entire Amazonian forest, so long as the gains from deforestation are invested in the construction of highways, factories, schools, and hospitals. The only condition of the sustainability of development, in this case, is that the sum total of capital (adding together artificial capital, human capital, social capital, and natural capital) does not diminish over time, but remains constant (see fig. 3.1).

If this condition is satisfied, future generations will inherit a quantity of capital equivalent to that available to present generations, even if the nature of the capital is modified. So long as one form of capital can be wholly substituted for another, the principle of sustainable development is not violated.

This model, known as the weak definition of sustainable development, is defended by the most optimistic analysts, who believe unreservedly in the power of technology. It seems unrealistic, however, for it ignores the irreversible character of certain ecological modifications. No amount of technological ingenuity will be able to replace the beauty of the Bengal tiger or the panda once they are extinct. The deforestation of the Amazonian forest in its entirety would have severe climatic consequences as well; in the best case, only a savanna with a low density of vegetation would be able to grow back again where today there is a forest rich in biodiversity.

The faith in perfect substitutability between different forms of capital also ignores the many functions performed by natural capital. A single forest furnishes wood and recreational space, provides a home for various flora and fauna, protects the soil against erosion, plays an important role in the carbon cycle, transpires water vapor that is then recycled in the form of rain, and so on. A highway, by contrast, permits the rapid circulation of a large number of vehicles transporting people and merchandise, but little else in the way of benefits.

Moreover, natural capital typically exhibits great diversity—the product of 3 billion years of biological evolution. The fauna that reside in a few square inches of forest soil are more varied than all the cars in the world put together. In nature, diversity is a major source of resilience in the face of external shocks: it is what allows the natural system to recover from such disturbances. The weak definition of sustainable development must therefore

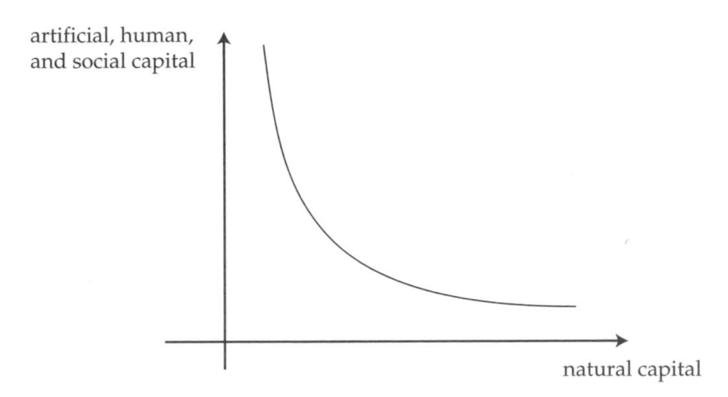

Figure 3.1 The perfect substitutability of capital.

be at least slightly amended to require that natural capital be maintained above a certain minimum level necessary for the maintenance of the essential services furnished by ecosystems.

The capacity to replace natural capital by artificial capital is a privilege of high-income societies and social classes. Drinking soda produced by industrial methods and sold in a bottle, because the local river water is polluted, is a technological alternative for the inhabitants of industrialized countries, but one that is denied to people in poor countries.

Zero Growth

An alternative form of sustainable development recognizes that natural capital and artificial capital cannot be wholly or perfectly substituted for each other. The condition of sustainable development is therefore modified to require that the quantity of natural capital not be modified over time by human activities, so that natural capital remains constant.

Economic development, of course, depends on the ability to substitute one resource for another. Indeed, one speaks of an industrial "metabolism" (thus, for example, coal, iron ore, and energy are transformed to produce steel). In order to prevent the stock of natural capital from being run down, the stock of artificial capital must be maintained at a constant level. On this view, defended by those who insist on the complete preservation of nature, sustainable development necessarily implies the absence of all economic expansion, or "zero growth."

This model, known as the strong definition of sustainable development, is socially unachievable. For the moral imperative of eliminating the extreme poverty from which a large part of the population of the world's developing countries suffers could then be carried out only through the redistribution of

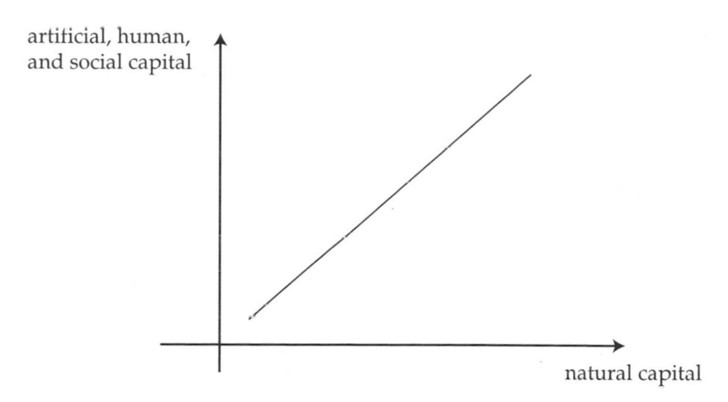

Figure 3.2 A "win-win" situation in which an increase in one type of capital is accompanied by an increase in the other.

existing wealth. This would lead to a significant reduction in the living standards of wealthy countries, which would therefore be disinclined to support a regime of zero growth.

Win-Win Situations

In certain situations, natural and artificial capital are complementary: an increase of one type of capital is accompanied by an increase of the other. These are sometimes called win-win situations, in which benefits in both environmental and socioeconomic terms are possible (fig. 3.2).

Agroforestry, a technique of agricultural production that reproduces the diversity of natural systems by mixing crops with tree cover and natural vegetation, exploits synergies between different plant species as well as between agricultural and forestry practices. For example, a stand of trees whose fruits and wood can be harvested casts shadow on the cultivated fields around them, which favors the growing of certain crops. Pasturelands can be alternated with these fields in order to produce organic manure. Various useful trees and shrubs can be planted as hedges along the fields in order to protect crops and soil from the wind and to provide a congenial habitat for certain animal species and pollinating insects. By integrating crop production, livestock farming, and forestry, such practices serve to increase the diversity of human-dominated ecosystems. The planting of certain kinds of fruit tree generates several of the goods and services furnished by a natural forest while providing nourishment as well as revenue from the harvesting and sale of their fruit. Ecotourism is another example of an activity where environmental protection goes hand in hand with economic development.

Examples of perfect win-win situations are rare, for the possibility can never be excluded that development will have unintended and unpredictable consequences, causing harm to certain compartments of an ecosystem. A hydraulic dam, for example, increases natural and artificial capital alike through the creation of a reservoir of water for domestic and industrial use, and through the production of energy, by allowing the adjacent farmland to be irrigated and controlling the downstream rate of flow of a watercourse. Secondary advantages from dams reside in new opportunities for fishing in the reservoir basin, the possibility of watering livestock, or the creation of recreational space.

But dams are also a source of ecological and social problems associated with the inundation of land upstream, which may be inhabited, or rich in forests or archaeological sites. The displacement of populations residing in the flooded area and the environmental impact of these populations on the territories where they resettle represent nontrivial social and ecological costs.

Other environmental effects of dam construction may be positive or negative depending on the situation and one's point of view. For example, because the water downstream is now poorer in silt and nutritive elements, irrigation canals will fill up less rapidly with sediments and treatment of water for drinking will be less expensive. A dam prevents catastrophic floods downstream that destroy crops and infrastructure, and assures a regular and predictable supply of water.

But the rich silt that seasonal floods used to deposit on farmland before the construction of the dam will have to be replaced by other methods of restoring the fertility of agricultural soils, notably the use of chemical fertilizers. The cost of this expedient may not be within the means of the poorest farmers. Fishermen in the estuary downstream from the dam will be confronted with a smaller population of fish, which will suffer from the water's reduced supply of nutritive elements and the dam's interference with their usual patterns of migration. There is a risk, too, that the distribution of natural vegetation downstream will be dramatically altered, which may have negative consequences for forestry, for example.

Experience shows that taking the environmental effects of dams into account at the outset of construction may, through sound management, maximize the positive effects and minimize the negative effects, both on the environment and on neighboring populations. Too often in the past, however, decisions about dam-building projects have been made with reference solely to economic considerations, with environmental consequences being considered only once the dam has gone into operation, and attempts made to remedy problems only once they are noticed. Today, even though an environmental impact assessment is part of every investment decision associated with dam construction, a few poorly conceived large-scale projects still manage to be completed.

Two important lessons can be drawn from the analysis of sustainable development in terms of the transformation of capital in its artificial, human, social, and natural forms. The first is that sustainable development requires a perpetual search for trade-offs between socioeconomic progress and the preservation of the environment. Extreme positions that insist either on a thoroughgoing substitution of artificial capital for natural capital or on prohibiting any decrease whatever in natural capital are simply not realistic. The only feasible alternative is to try to achieve an optimal level of transformation of natural capital into artificial capital that will satisfy the aspirations of human societies while preserving the vital goods and services provided to them by ecosystems.

The second lesson concerns those situations in which economic development is compatible with maintaining, and indeed improving, the function of ecosystems. This type of sustainable development requires taking into account all the relevant environmental factors, and undertaking an exhaustive scientific analysis of the ecological processes involved. In the search for an optimal balance between economic and environmental costs and benefits, there is no ready-made solution: each situation is different. Nonetheless, there is much to be learned from errors and successes alike. The essential thing is to determine the optimum level of use for a given store of natural capital.

Second Model: Carrying Capacity

The concept of carrying capacity represents the limit imposed by the natural environment on the population that uses the goods and services furnished by the environment, and measures the maximum population of a given species that an ecosystem can feed in a sustainable manner. By this it is meant that the population must be able to occupy a habitat for a long period of time without reducing the habitat's capacity to support an equivalent population in the future.

This notion was originally developed by managers of grazing lands to estimate the number of head of cattle that could profitably be kept per acre. The term "ecological carrying capacity" is used to refer to the maximum number of animals that can be grazed per acre without degrading the land, and "economic carrying capacity" to refer to the number of animals for which the level of productivity in milk or meat per head is highest. Because this second carrying capacity is smaller than the first, the economic optimum implies a lower density of cattle than the ecological maximum that can be sustainably supported.

Extended to the human population, the concept of carrying capacity describes the maximum population that can satisfy its needs in a sustainable way on the basis of the flow of goods and services that the natural capital of a region, or of the earth as a whole, is capable of generating. The many estimates of the size of the human population that the planet can sustainably support vary, depending on the assumptions their authors make, between 7.7 and 12 billion people. By way of comparison, the size of the world population in 2050 is plausibly projected to be in the neighborhood of 9 billion. Although both these types of reckoning suffer from a great many uncertainties and approximations, they nonetheless suggest that in the coming decades humanity could approach the planet's carrying capacity.

Researchers also estimate that 40 percent of the biological production of the planet's terrestrial ecosystems has already been appropriated by the human race for its own consumption. Studies of the extent of mankind's "ecological footprint"—that is, the area of productive land and oceans needed to produce the resources consumed by humanity and absorb the waste generated by it—estimate that already by the early 1980s this footprint exceeded the total area of the planet. In 2000 this surplus area was calculated to be on the order of 20 percent. If indeed humanity is running down its supply of natural capital, rather than living on the interest that it produces, such calculations would appear to support a pessimistic view of the environmental outlook.

Optimists are not lacking for arguments in challenging crude estimates of this sort, however. While applying the concept of carrying capacity to an animal population poses no particular problem, extending it to the human species, whether in relation to a specific ecosystem or on a planetary scale, is more difficult. Unlike the situation with herds of cattle or sheep, resource consumption differs among groups of human beings, and changes over time in respect of both quantity and composition. Human societies constantly innovate: new technologies of production are developed, more sustainable practices for managing the environment are put into effect, social institutions that regulate access to resources are established, and the preferences of consumers evolve. International trade and migration continually rearrange patterns of production and consumption as well. Taken together, these various innovations broaden the limits that ecosystems place upon the growth of the human population and increase the carrying capacity of the planet. The concept of carrying capacity must therefore take into account the consumption and technology of a given population.

Third Model: The Ecological Impact of a Society

In this model, the factors overlooked by the concept of carrying capacity are taken into account in order to provide a simple representation of the ecological impact of a society upon its environment. This impact is assumed to be the product of multiplying the number of persons using an ecosystem first by the average per capita consumption of resources, and then by the environmental impact of the technologies used to produce the goods consumed: Impact = Population × Affluence × Technology (or I – PAT, as this formula is generally known).

If one follows Barry Commoner in applying this formula to the ecological impact of beer consumption, one obtains:

> I (number of empty beer bottles thrown away) =
> P (number of inhabitants) × A (quarts of beer consumed per inhabitant) × T (number of bottles per quart of beer)

The variables "number of inhabitants" and "quarts of beer" cancel out, since each one appears once in the numerator and once in the denominator to the right of the equal sign, leaving only the variable "number of bottles" on either side. (We are left, then, with an identity, which is to say a mathematical expression in which the same term is found on the left and right of the equal sign).

The principal virtue of this identity is that it clarifies the nature of the interaction between certain factors that lead to an impact of human activity on the environment. Although human population growth is often claimed to be the chief cause of the degradation of the environment, this model brings out the fact that the consumption of each individual must be taken into account in order to estimate per capita environmental impact.

Thus, although demographic growth in Burkina Faso is three times higher than in the United States, an inhabitant of Burkina Faso consumes on average roughly fifty to one hundred times less energy than an American. It is also necessary to multiply these two terms by the environmental impact of the production technologies employed. The hoe that the inhabitant of Burkina Faso uses to cultivate his field and the donkey that carries his harvest to the local market have considerably less environmental impact than the factories and intercontinental transport system that support food production and distribution in developed countries, not to mention the sport utility vehicles used by American families to do their shopping.

While the earth's population grew fourfold between 1890 and 1990, energy production increased worldwide by a factor of sixteen over the same

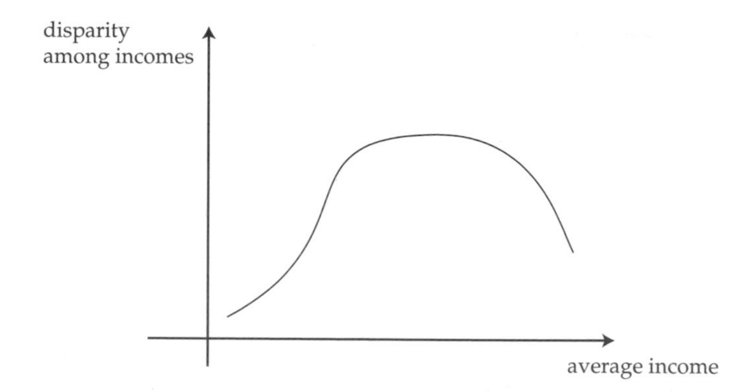

Figure 3.3 Kuznets curve demonstrating the relationship between average income and the disparity among incomes.

period. This suggests that the consumption and technology factors in the I = PAT equation grew substantially during the course of the last century. Technology has an ambiguous role in this model, however. To be sure, technological progress has brought with it the pollution of coal-burning factories and individual cars; but it has also, as the optimists rightly remind us, led to many cleaner and less energy-consuming technologies, thanks to which the air in the cities of the world's wealthy countries is once again fit to breathe. This ambivalence is taken into account in the following model.

Fourth Model: The Environmental Kuznets Curve

The American economist and Nobel Prize winner Simon Kuznets suggested in 1955 that the disparity between the incomes of a country's population initially increases with economic development, but then decreases once per capita income rises above a certain threshold. The relation between average income and disparity among incomes may therefore be represented as a bell-shaped curve (fig. 3.3).

Although the limits of this Kuznets curve, as it is known, have since been demonstrated, it continues to influence debates over the role of economic growth in the reduction of poverty. It has also given its name to another curve that is of greater interest for our purposes here, and which has in common with the first curve only its bell shape. By virtue of this similarity of form, it is known as the environmental Kuznets curve (fig. 3.4).

This curve shows an increase in the environmental impact of a society in the early stages of economic development, followed by a lessening of this impact once per capita income has exceeded a certain threshold.

The environmental Kuznets curve predicts that countries that pollute the

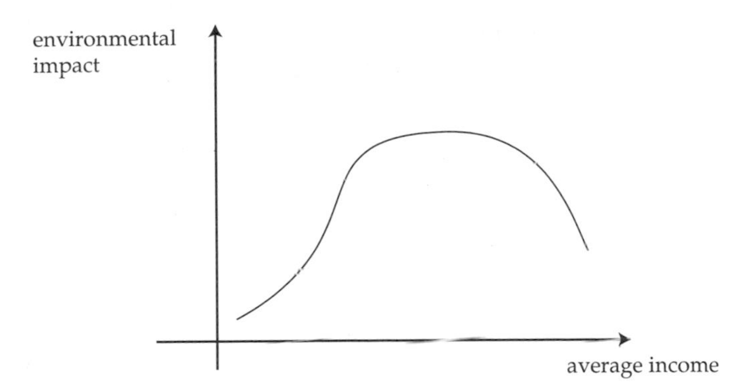

Figure 3.4 The environmental Kuznets curve.

least will be both the world's poorest, which have neither factories nor a transportation system that would enable their citizens to have a great impact on the environment, and the world's richest countries, which have acquired the means to develop clean technologies. By contrast, the countries with the highest environmental impact should be those whose economies are in transition.

Although this model appears to support the claims of the optimists, since a decline in pollution is due to the technological progress generated by economic growth, pessimists find in it support for their position as well. For not only are developing countries, which today represent 80 percent of the world's population, now entering a phase of high pollution and economic growth, further growth will cause the level of pollution to rapidly increase in the years ahead. Surely, say the pessimists, if the earth's climate and ecosystems, terrestrial and marine alike, were so profoundly modified during the nineteenth and twentieth centuries when only 20 percent of its population (the earliest countries to industrialize) had "dirty" economies, the other 80 percent (developing countries) having "clean" economies, it has little chance of withstanding the crisis that will occur once these percentages are inverted during the twenty-first century.

The Balance of Evidence

Is the model associated with the environmental Kuznets curve confirmed by observations? Only a few forms of local pollution that pose a short-term problem clearly exhibit a bell shape as a function of average income in a given country. One example is air pollution in urban areas, in particular pollution due to sulfur dioxide (SO_2) and carbon monoxide (CO). These forms have declined by 66 percent between 1976 and 1998 in the United States, thanks to technologies that emit less pollution per unit of production. The contamination of drinking water also conforms to this model.

By contrast, carbon dioxide (CO_2) emissions, responsible on a global scale and over the long term for the strengthening of the greenhouse effect and global climatic warming, continue to increase with a country's income. It may be that an inflection point will appear eventually, and that at some point in the future this type of environmental change will likewise display an environmental Kuznets curve. But there is no reason to believe that this inflection point will appear before a major and irreversible modification of climatic conditions occurs, the socioeconomic impact of which is likely to be very costly indeed.

Let us turn to the evidence regarding deforestation. In certain countries a tendency to deforestation is observed over a period of decades as a consequence of agricultural expansion in the early stages of economic development, followed by a period in which the forest cover stabilizes as a result of an increase in agricultural yields rather than of the area under cultivation. Next, farmland is abandoned and reforestation occurs as the economy industrializes. This forest transition, though it is consistent with the environmental Kuznets curve, by no means occurs everywhere, however. What we have here, then, is a contingent model that describes a possible outcome under certain conditions.

The diminution of certain forms of environmental degradation in advanced economies is not a spontaneous effect of economic growth. It is mainly the consequence of vigorous environmental policies, but also of the development and application of less polluting technologies. The adoption of such policies and technologies depends on several things: the orientation of consumption toward ecological values, once material needs have been satisfied; consumers' ability to pay for environment-friendly goods and services; the action of interest groups on behalf of environmental protection; and the ability of governments to draft strict environmental legislation and enforce compliance with it.

Reduced environmental degradation is also to some extent the result of transferring polluting industries or toxic waste from wealthy countries to poor ones. For example, more than 50 percent of obsolete consumer electronics items in the United States, which are laced with lead, cadmium, and other toxic materials, are being sent to nations such as China and India for unregulated disposal. This and other facts make it difficult to conclude that economic growth is beneficial to the global environment, since the problem is routinely being exported to parts of the world that are least prepared to tackle it. Some of the most polluting activities, however, such as the production of electricity and the local transport of people and merchandise, cannot be moved to remote locations, and so do not participate in this transfer of pollution to the margins of the advanced economic world.

Generally speaking, a tendency to environmental degradation can be reversed only on two conditions. The first is that the negative effects of degradation on a particular population must be perceived by the population itself and by its political leaders. The second is that the persons who bear the cost of the measures aimed at reversing the damage must be the same ones who profit from the success of such measures. This explains why the environmental Kuznets curve is found to obtain only for local environmental problems with a quick recovery time such as urban air pollution and the contamination of drinking water, and not in the case of global warming or the planet's loss of biological diversity, where the beneficiaries of a lessening of environmental degradation would be future generations, everywhere in the world, rather than the people who make sacrifices today in a particular place or region.

The model implied by the environmental Kuznets curve introduces the first of several nonlinear relations characterizing the interaction between human activity and the natural environment: economic growth, if it is to be able to solve environmental problems, must be accompanied not only by institutional and cultural changes, but also by policy intervention.

Fifth Model: The Law of Diminishing Returns and Surpluses

A second nonlinear relation needs to be introduced at this juncture. Let us imagine a rapidly growing population of farmers living on a fixed area of land. Assuming that per capita consumption remains constant, the growth of the population is accompanied by a linear increase in total consumption: a doubling of the population leads to a doubling of consumption, and so on.

But a human being is not merely a mouth needing to be fed; each one also has hands for working. Assuming that the technology employed remains unchanged, the growth of the population is accompanied by a nonlinear increase in production (see fig. 3.5).

The shape of the curve reflects the law of diminishing returns, also called the law of diminishing marginal productivity. In a process of production, the marginal product of a factor of production (known as an input) is defined by the increase in the total product due to the increase of a unit of input. The law of diminishing marginal productivity says that, beyond a certain threshold, an increase of inputs no longer yields a proportional increase in production: the marginal product diminishes and may even, in the extreme case, become negative.

Suppose, for example, that a heavy piece of furniture has to be brought

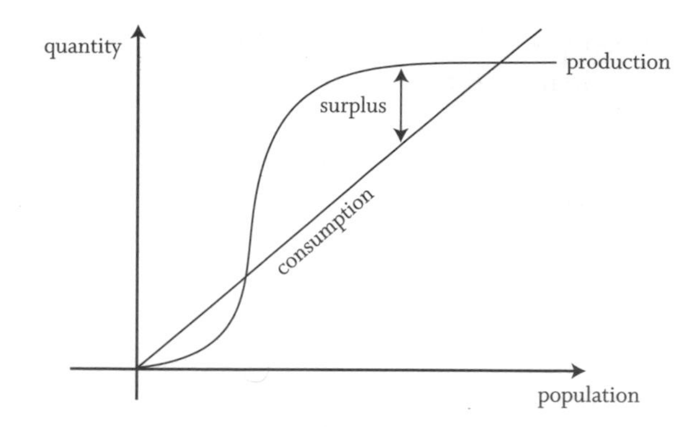

Figure 3.5 The law of diminishing returns.

down a staircase. One man doing the job alone would have a hard time of it. If one or two others lend him a hand, progress will be faster. If the job is assigned to between four and six movers, the ideal number, a threshold will have been reached: if any more people are added to the team, the gain in productivity per mover will diminish, because the movers will keep getting in each other's way. There will even come a point when adding further movers will make the task impossible: the staircase will be so crowded that the piece of furniture cannot get through.

It is easy to transpose this example to agriculture. A shortage of labor prevents a high level of production from being attained. When the optimum number of farmworkers is employed, however, it becomes possible to cultivate the most fertile land and jointly achieve a high level of production. Too great a number of workers forces a fraction of the farming population to cultivate marginal land with low yields, in which case productivity will be small and the risk of degradation great.

The Fate of Surpluses

Given the linear form of the consumption curve and the nonlinear form of the production curve, one observes in figure 3.5 a period during which agricultural production exceeds the needs of the farming population. The use made of this surplus to a large extent determines the environmental impact of a society in the near future.

Ideally, the surplus will be invested in research, experimentation, and the development of new production technologies in anticipation of a reduction in marginal product associated with population growth. This process of innovation requires an elaborate system of social organization, since those who are engaged in research and development, by definition a risky activity, depend for their survival on the surplus generated by others. Technological innovation makes it possible to avoid overexploitation of the land, and the food shortages and ecological crises that accompany it, once the marginal product has become low or null. A new and more effective production technology will be associated with a new production curve that, initially, gives rise to a high and stable marginal product.

If the surplus is absorbed by expenditures aimed at enhancing the prestige of a local elite or by military expenditures that strengthen its position, or if the surplus is diverted from the local economy by a colonial power, or if it is consumed in order to survive a natural catastrophe (a prolonged drought, for example), farmers will no longer have the means to invest in technological innovation. In this case, a lack of capital will prevent the necessary adjustments from being made, with the result that a growing population finds itself caught up in a vicious circle: as more and more farmers live in a limited area, using the same agricultural techniques, the land becomes progressively more degraded—to the point that marginal yields become negative. The American anthropologist Clifford Geertz coined the term "involution" for this spiral, in which environmental degradation and malnutrition reinforce each other.

Archaeological evidence reveals that this process has led to the collapse of complex societies and civilizations in the past. Easter Island, with a population of seven thousand around 1400 A.D., subsequently experienced complete deforestation, severe erosion of the soil, and the virtual extinction of local species of seabirds and shellfish. A sophisticated society of Polynesian origin that had sculpted in stone some nine hundred giant statues, two hundred of which were arranged in precise astronomical alignment, was reduced to a handful of primitive and ignorant cannibals when the Dutch explorer Jacob Roggeveen discovered the island in 1722.

Archaeologists have succeeded in showing that this society squandered its surplus in human capital (a peasant labor force) and its natural capital (*Jubaea*, a species of large palm tree) in competitions among rival clans for prestige, symbolized by the erection of ever-larger statues. No influential member of this society appears to have foreseen the ecological disaster to which an economy based on such displays of ostentation would lead the island and its inhabitants, despite the sophistication of the culture and the prosperity of the economy that had grown up there. Once the last tree was cut down, what was once a race of able sailors, who around 400 A.D. had crossed a part of the Pacific Ocean against prevailing currents and winds, no longer had the means even to construct a boat in order to fish or to leave the island.

Other cases, less extreme but no less telling, may be cited. The decline of Mesopotamian and Hohokam civilizations, for example, was due in part to overly intensive agriculture, which led to both salinization of the soil and deforestation, as well as to excessive fishing and hunting of game. While not directly responsible for the collapse of these societies, these activities decisively increased their vulnerability to repeated droughts and to both internal and external sociopolitical upheavals.

In all these cases, an entrenched political elite diverted a growing part of the surplus generated by the peasant class for the purpose of enhancing its prestige or of undertaking military campaigns. Environmental pessimists fear the same syndrome may be reproducing itself today, only in a modern form and on a planetary scale.

Innovation and Scarcity

Optimists, by contrast, are able to point to many episodes in human history in which ecological stress and a scarcity of certain vital resources have proved to be a source of great inventiveness.

In *Poverty and Progress* (1973), the British economic historian Richard Wilkinson argued that the principal stimulus of economic development has always been an increase of material needs beyond the maximum level that could be satisfied by the environment. The great technological revolutions were often preceded by ecological crises. Analyzing the scarcity of land and timber in England before the Industrial Revolution, Wilkinson observed that overexploitation of forests, and the growing shortage of fuel that resulted from this, led to a reliance upon coal—the point of departure for a series of technological innovations, among them the steam engine, which was first used to pump water out of coal mines.

A few years earlier, the Danish economist Ester Boserup had advanced a similar argument with respect to agriculture in *The Conditions of Agricultural Growth* (1965). In several places in the world—in Europe during the agricultural revolution of the eighteenth century and in the course of the twentieth century in densely populated regions of Asia—an increase in crop yields and a shortening of the period during which farmland lay fallow (the method traditionally used to restore its fertility) were stimulated by an increase of the population beyond what the environment could support. Farmers were forced to increase agricultural inputs (organic and chemical fertilizers, for example) in order to reduce the area of land that lay fallow without exhausting the soil. Intensifying agriculture in this manner required capital investment and, prior to the introduction of farming machinery, an increase in the amount of labor per unit of production. The successive thresholds of agricultural intensification achieved by the application of new technologies,

Boserup argued, correspond to periods when the carrying capacity of an environment has been reached. Ecological crises are overcome, then, because human societies invest in research to develop advanced technologies when they have scarcely any choice left on the ecological level.

In sum, the course of environmental degradation does not depend solely on the size of a population, its level of consumption, and its existing technology. It also crucially depends on a society's success or failure in developing new technologies that make it possible to anticipate imminent pressures. Technological innovation requires that a certain surplus of capital and labor be available. Beyond this, success in avoiding environmental degradation depends on a society's ability to detect beforehand the impact of human activity on the environment; the rapid and direct transmission of information to policy-making centers; a recognition that the state of the environment is the product of complex systemic processes that need to be understood; and the creation of consensus among different social groups, so that a minority elite is not able to enrich itself at the expense of the environment, and therefore at the expense of the majority of the population.

All of this means that a society must develop and maintain institutions that are capable of responding rapidly, and effectively, to pressures upon environmental resources.

Sixth Model: The Tragedy of the Commons

One of the ways in which a society's institutions influence the interaction between human activity and its environment is by regulating access to natural resources. The American ecologist Garrett Hardin, in a famous article published in 1968, drew a parallel with the life of a small rural community of the sort that existed in Europe during the Middle Ages. Each member of the community farms a few fields and owns a few sheep. All of the sheep in the community are led out each day to graze. While the sheep belong to individual farmers, the pastureland is held in common. The quality of this land, poor to begin with, will be rapidly worn out if the herd grazes on it too frequently and too intensively.

A self-seeking farmer will be tempted to try to increase the number of sheep he owns. Each additional sheep brings him extra income in the form of more wool and lambs each year. This costs him nothing, for the additional sheep feed on land that belongs to the community. The moment will come, however, when increasing the number of sheep beyond a certain point leads to the impoverishment of the pastureland. The consequences are distributed more or less equally over the sheep of the village, which all suffer to some small degree from the reduction in quantity and quality of the grass available to them. The additional sheep therefore bear only a small fraction of the ecological cost for which they are responsible. The villager who acquires more sheep therefore stands to benefit since his gains are greater than the cost he incurs as a result of the minimal harm done to the health of his herd. Indeed, he can go on adding to his herd until the community's grazing lands are worn out. But at this point his sheep-breeding business will have lost all profitability, dragging down the whole community with him. The process of degradation will occur all the more quickly as other villagers are tempted to imitate their neighbor in the hope of enriching themselves at the expense of land belonging to the community.

This situation—known as the tragedy of the commons, after the title of Hardin's article—describes a conflict between individual and collective rationality, which is to say between the private and the public good. The same mechanism is responsible for the pollution of the atmosphere and the decline of the stock of fish in the oceans. It seems to argue in favor of the pessimists.

When we use common-pool resources, whether to make things from them or to dump waste in them, our behavior inevitably has indirect consequences for other users of these resources: they do not pay for activities from which they benefit, nor are they indemnified in the event they are adversely affected. Such consequences are called "externalities."

Externalities may therefore be either positive or negative. When a large number of countries ratify and put into effect an international agreement to lower greenhouse gas emissions responsible for climate change, while one country refuses to ratify the convention, this is an example of a positive externality: all countries benefit from it, including the one that is not a signatory to it. By contrast, a country that overfishes in international waters, exhausting the stock of certain species at the expense of other countries' fishing fleets, is the source of a negative externality.

Some scientists claim to find a general law in the fact that individual users of the environment neglect negative externalities. The truth of the matter, however, is that this syndrome of degradation appears only under quite particular circumstances. It is necessary, first, that the natural resource in question be intrinsically common in nature. Resources whose physical properties or behavior makes access by potential users difficult to control—aquifers or migratory animal species such as whales, for example—come under this head.

The second distinguishing characteristic of common-pool resources is that use by some is liable to diminish the welfare of other users of this resource, either because the resource is limited in quantity or because it is renewed only slowly. If one person extracts a large quantity of water every day from a well, the level of water in adjacent wells falls and the cost of extracting it increases for neighboring users. High exclusion costs and a reduction in general welfare are therefore typical consequences.

Finally, a tragedy of the commons occurs only where access to resources is free for everyone. As a practical matter, access to a particular resource is often controlled by laws regulating the use of private, community, or national property. Such laws, optimists insist, make it possible to exclude nonauthorized users and protect the resource from abusive exploitation by authorized users.

State, Private, and Community Property

Throughout history, human societies have displayed considerable ingenuity in devising institutional arrangements that fulfill these twin functions of exclusion and regulation. For example, hunting and the harvesting of forest products can be regulated by restricting periods of activity and the kinds of technology that may be used. When exclusive fishing rights in coastal waters are awarded to fishery cooperatives, these groups limit the volume of fish that may be taken by each member and sometimes mandate the sharing of revenues, in order to discourage some members from fishing more than others.

In semi-arid regions in Arab countries of the Near East, allowing only members of a particular tribe to have access to wells makes it possible to economically control the use of vast grazing lands surrounding these wells. In the United States, the rate of oil drilling by small producers in Wyoming was slowed by a legal decision assigning private property rights not to the oil actually extracted, but to units of oil underground, prior to its extraction—thus removing the incentive to pump as much oil as possible before a competitor came along and reduced the supply. In certain Japanese villages, thatch used to be harvested collectively and the bundles allocated by lot to each family. In parts of Asia, water is distributed among the various irrigated plots of farmland in a village in accordance with precise rules enforced by water management associations. Thus, for example, the fields upstream can be irrigated only once those downstream, which are exposed to the greatest risk in the event of a drought, have received the amount of water they need.

At the risk of overgeneralizing, community property regimes appear to be the most effective means of regulating the use of resources that are common by their very nature. The differential effects of alternative regimes are illustrated by a famous satellite image, taken in 1989, of grazing lands in central Asia, covering parts of China, Russia, and Mongolia: whereas three-quarters of the Russian zone and a third of the Chinese zone display signs of severe degradation, only a tenth of the Mongol territory shows such signs.

In Russia the pasturelands were the property of the state and managed by nationalized agricultural enterprises. This form of management involved the concentration of a large number of cattle each year in fenced-off meadows, which led to overgrazing and degradation from the continual trampling of the ground by the animals. Meanwhile, in adjacent fields devoted to growing fodder for the cattle, the repeated use of heavy agricultural machinery destroyed fragile soils. In China pasturelands that had been collectivized in the 1950s were put back into private hands at the end of the 1970s, and the territory divided up among a great number of individual owners, preventing the cattle from following their old migration routes. In Mongolia, by contrast, communal management of grazing lands remained in force, assuring the mobility of herds across the vast steppes in response to seasonal climatic fluctuations and in accordance with ancestral customs.

In this case, both state and private property regimes led to severe degradation, whereas a traditional community regime for managing a natural resource preserved the herdsmen's capital. The opportunistic and mobile method of pasture management practiced by the Mongol tribes nonetheless bears little resemblance to a true regime of free access, since it involves strict rules of cooperation and reciprocity.

Historically, access to state-owned property has often effectively been free. Nationalized resources suffer from difficulties in enforcing rules of use and monitoring compliance with them, owing to bureaucratic inefficiency and inadequate budgets, with the result that the imagined benefits of public ownership are rarely achieved. In the absence of proper enforcement, ad hoc laws and rulings proliferate, further encouraging noncompliance. Nationalization has seldom improved the management of natural resources, particularly in parts of the world where state structures are weak.

With regard to private property regimes, the costs associated with the exclusion of unauthorized users and the protection of the privatized resource can be heavy for owners, particularly in the case of common-pool resources. In the Middle Ages in Europe, seigneurial forests were frequently subject to poaching and illegal takings by the starving peasantry: inequalities in the distribution of rights of access to resources made respect for private forest property impossible to enforce. In Africa today small businesses that charge tourists a fee for the right to view white rhinoceroses, an endangered species, in cordoned-off areas have had to install a costly infrastructure of electric fences and armed guards to protect the animals from poachers.

As a matter of economic profitability, it is generally optimal to harvest late-maturing resources (whales, for example, or trees such as the sequoia, whose growth cycle may be as much as two thousand years) until they are exhausted, rather than to exploit them in a sustainable fashion. Only by taking into account the interests of other potential users and future generations is it possible to justify restraint in the exploitation of this type of resource. Altruistic considerations of this sort are not generally compatible with a regime of private property, however.

The management of resources by small communities of interdependent users, on the other hand, has often helped to prevent ecosystems from being overused, so long as these communities have been granted the legal right of exclusive use of the resources in question. This is true of fishing communities that depend on regularly replenished stocks of fish or shellfish near their shores, whether in Japan, certain islands in the Pacific, or off the coast of Maine, in the United States, where lobster has (despite occasional intervals of scarcity) long been abundant. Because some of the medieval rural communities mentioned by Garrett Hardin decreed limits on the number of head of livestock that each villager could own, their grazing lands survived for several centuries without being degraded.

The capacity of human societies to establish institutional forms of cooperation between interdependent users of natural resources, and to lay down and adjust norms of behavior, is an important factor in preventing the degradation of common pool resources. Where they exist, such institutions make it possible to avoid the destructive impact on the natural environment that arises from a divergence between individual and collective rationality.

They may be absent, however, when an environmental problem has only recently emerged, as in the case of global warming until a little more than a decade ago, for example, or when traditional modes of resource management have been swept away by social and political upheavals. In the latter case, the dislocation of pastoral societies by modernization often leads to a weakening of the customary systems of controlling resource access and use, thus degrading the environment. Resources whose use had been strictly regulated by local institutions, sometimes complex and informally structured, then fall into a de facto regime of free access. If the demand for these resources is high, they end up being overexploited.

Regimes that control access to natural resources and regulate their use may also lead to a degradation of the environment if they do not change over time. A society must preserve its ability to rapidly and efficiently innovate when new challenges to its institutions arise. Two notable examples of institutional innovation on a global scale are the Montreal Protocol on Substances That Deplete the Ozone Layer, which was signed in 1987 and whose positive effects are now beginning to make themselves felt; and, more recently, the Kyoto Protocol, intended to combat climate change. The second of these protocols has been criticized, with some justice, for responding too slowly and too timidly to a major environmental challenge. But it needs

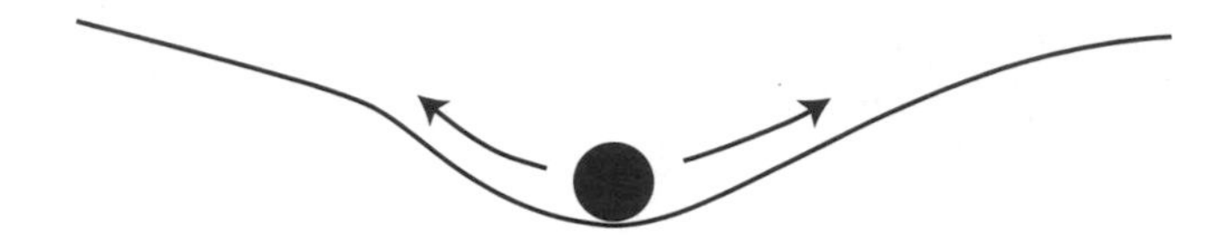

Figure 3.6 A valley that is both wide and deep represents an unstable but very resilient state.

to be kept in mind that it is more difficult to develop institutions for managing global resources, which to varying degrees affect the whole of the present and future world population, than it is to develop local institutions for managing resources confined to a particular ecosystem and used by a single community.

Seventh Model: Stability vs. Resilience

The fragility (or, by contrast, the resistance) of an ecosystem is a measure of stability, which is to say the system's capacity to withstand the effects of external stress and disturbance. A fragile ecosystem is unstable. A resistant ecosystem, on the other hand, is stable: it remains in the same state no matter what perturbations it is exposed to.

The concept of resilience, by contrast, describes the capacity of an ecosystem to recover from a shock and return to its initial state. Resilience can be pictured in terms of a marble rolling over an undulating surface. A valley that is both wide and deep represents an unstable but very resilient state: the marble is easily displaced but rapidly comes back to its initial position when the disturbance ceases (see fig. 3.6). Like a reed that bends in the wind but then stands up again, resilience depends on mechanisms for absorbing shocks.

A narrow, relatively shallow valley, on the other hand, represents a stable but unresilient state. Under normal circumstances, the marble does not move. When it is displaced by a large external force, however, it risks never returning to its initial position; instead, like a tree lying flat on the ground after having been uprooted by a storm, it occupies a new stable state (see fig. 3.7).

An ecosystem's degree of resilience is often a better indicator of its "health" than its stability. A stable system is often unresilient because it has rigidly protected itself against minor disturbances, rather than develop mechanisms for flexibly coping with major disturbances. For example, industrialized countries have constructed vast networks of sea walls and dams to protect their alluvial plains and large cities against inundation. This has made it necessary to replace the natural fertilization of the soil by flooding with sizable amounts of chemical fertilizer. Under exceptional circum-

Figure 3.7 A narrow, relatively shallow valley, on the other hand, represents a stable but unresilient state.

stances, however, when floodwaters rise to such a height that the barriers give way, the economic cost (in terms of the destruction of crops and infrastructure) is considerable: the system has traded resilience in the face of extreme conditions for stability under normal conditions. In the wake of hurricanes Katrina and Rita, which severely damaged the Gulf Coast of the United States in 2005, restoration projects have therefore been proposed that would combine strengthened levees and a web of self-sustaining barrier islands, wetlands, and coastal forests capable of acting as buffers against storm surge.

By contrast, Bangladesh, one of the poorest countries in the world, has learned to live with floods. Both habitat and agriculture are adapted to frequent inundation. The houses have flat thatch roofs on which families are able to take refuge with their belongings in the case of exceptional flooding. If necessary, these roofs can be detached from the walls and floated like rafts. Once the waters have subsided, a family is able to reinstall the roof-raft atop the walls of the house and carry on with its life. In Bangladesh the highest agricultural yields are obtained just after a flood, since the fields are enriched by the silt deposited by rivers that have overflowed their banks. The species of rice cultivated in the floodplain are highly flood-tolerant owing to their long stalks and sturdy roots.

An example of the same sort is the excessive use of antibiotics. These bring benefits to the sick in the short term but render the human population less resilient in the face of new epidemics caused by antibiotic-resistant pathogens. Such resistant strains are the consequence of natural selection among pathogenic agents, itself accelerated by the massive worldwide use of antibiotics.

The concept of resilience applies to socioeconomic as well as ecological systems. Assessing the resilience of an ecosystem strongly dominated by human activity to both natural and socioeconomic disturbances is sometimes difficult. It is now possible to determine, for example, which system of rangeland management is best suited, in the face of episodic droughts, to maintain the resilience of a semi-arid ecosystem that supports a large population of livestock. It is also possible to determine, on the basis of socioeconomic data, whether an agricultural enterprise will be resilient in the face of erratic fluctuations in the price of meat or of state subsidies for dairy production. It is more important, however, to be able to determine whether the integrated system formed by semi-arid grazing lands and the farms that exploit them will be resilient in the face of a drought that coincides with a drop in the market value of meat and an abandonment of dairy subvention policies. This type of analysis requires a detailed understanding of the interactions between ecological and socioeconomic systems.

Innate Resilience and Human Intervention

Resilience depends on two sets of factors: on the one hand, diversity, redundancy, and the heterogeneity of the elements that make up the system; on the other, the existence of built-in mechanisms for reorganizing the system in the event of disturbance.

Biodiversity is one of the chief elements responsible for the resilience of natural ecosystems. In normal times, only a small number of species play an important role in the functioning of ecosystems. Many species are superfluous. If some are eliminated by an external shock or by an epidemic, others take over their function. (For the same reason, crucial pieces of equipment in spacecraft are duplicated in order to have a reserve supply available in case of failure.) Moreover, when an ecosystem suffers a disturbance, a great number of species that are not useful in normal times act as buffers, permitting the ecosystem to maintain its vital functions. In this case, both the diversification and complementarity of species are essential.

Peasants in Sahelian Africa follow the same strategy. In the event of drought, low crop yields force them to produce more meat and milk to survive. Scarcity causes the price of cereals to rise. Many peasants then sell their surplus cattle and the price of meat falls. With the return to normal weather conditions, the price of cereals falls, permitting farmers to replenish their stocks, while the price of cattle rises, providing an important source of revenue for those who made sacrifices in order to keep their cattle during the drought. Diversity therefore represents a form of insurance against shocks. It also provides the means necessary for reorganizing an ecosystem or a way of life after a severe shock.

The second factor that creates resilience in a system is the existence of mechanisms—negative feedback loops—that help to bring about its rapid reorganization. When a disturbance affects the system, these loops quickly get the system back on track.

The example of forest fires in northerly latitudes (and sometimes, as in the case of eucalyptus forests, in temperate zones) is instructive in this regard. Leaves and dead branches accumulate on the ground, creating fuel for potentially destructive fires, which are apt to spread by lightning. The best way to reduce the risk of such destruction is to dispose of this fuel.

Under natural conditions, in some forests small frequent fires accomplish this function. If the fuel does not have the opportunity to accumulate in great quantity, the fires do not reach high temperatures. Dry plant matter burns on the ground, damaging some, but by no means all, trees in the vicinity. In this way a kind of natural selection is carried out, with the result that only the most fire-resistant trees continue to grow. Moreover, the seeds of certain plants germinate only if they are "stimulated" by the heat of a fire. Young trees grow back more rapidly after a fire because they profit from the fertilization of the soil by the ashes. Successive small fires also create a mosaic of vegetation at different stages of growth, producing a fragmented and diverse landscape, so that the next fire will spread in a limited way owing to spatial discontinuities in the quantity and type of available fuel. Finally, this fragmentation of the landscape guarantees a rapid regeneration of burned areas, for a stock of seeds is maintained in the neighboring areas spared by the fire.

Taken together, these processes constitute a negative feedback loop that permits the forest to bounce back from small frequent fires. Under natural conditions, fires are part of the life of certain forests. In the absence of fires, fuel accumulates on the ground, species of trees develop that are not resistant to fire, and the forest's vulnerability to an exceptional, large fire event increases. In this case the high temperatures produced by the burning of a large quantity of accumulated plant matter on the ground would have a destructive impact on the trees and soil.

Human intervention, in the form of land management policies, can substantially modify the resilience of ecological and human systems. Subsidies for intensive farming, for example, which have reduced the diversity of the agrarian landscape, are a source of concern for environmental pessimists: the sort of intensive monoculture practiced today is more vulnerable than the mixed farming of earlier times to variations in world market prices, the degradation of soils, and seasonal crop diseases.

Well-intentioned but naive policies intended to bolster the stability of ecosystems have also worked to eliminate natural regulatory mechanisms. A misguided desire to "protect" forests against fires at the beginning of the twentieth century in the United States, for example, led to costly and counterproductive attempts to put out every single fire, no matter how small, at once. Fuel accumulated on the forest floor for several decades, after which devastating fires ravaged the landscape for twenty years. Following a catastrophic fire that destroyed almost half of the vegetation of Yellowstone

National Park in the late 1980s, forest managers discovered the virtues of frequent small natural fires.

Farmers in the savanna regions of Africa have long applied a similar principle. They burn vegetation in a controlled manner early in the dry season, while it is still damp and does not rapidly catch fire. This intentional fragmentation of the landscape by means of controllable fires that are confined to small, dispersed patches prevents accidental fires from spreading over large areas later in the season, when the vegetation is dry.

Environmental optimists look to enlightened human intervention as a way of increasing the resilience of ecosystems dominated by human activity. It is possible, on the one hand, to increase the adaptive capacity of the key actors of a system through institutions that promote learning, flexibility, and a willingness to respond rapidly to new challenges with fresh solutions. All these things argue in favor of an adaptive approach to managing ecosystems. On the other hand, the introduction of artificial negative feedbacks through new institutional arrangements improves the system's capacity for reorganization. The creation of markets for natural resources is an example of this type of mechanism: when a disturbance degrades a resource, its scarcity causes its market price to rise, which encourages users to search for less costly substitutes, and therefore to reorganize the system of production.

The way in which human societies manage the natural environment therefore has a considerable impact on the resilience of ecosystems in the face of disturbances. Resource management emphasizing stability and efficiency in order to satisfy short-term needs may diminish the long-term resilience of the ecosystem. Over time the institutions that manage resources become more rigid and less sensitive to the signals sent by the environment. The natural mechanisms that give the ecosystem its resilience are weakened, and the ecosystem is kept in an artificially stable state by controlling the impact of natural factors of variability. When an extreme natural phenomenon occurs—a catastrophic flood, a major forest fire, an epidemic, or a food shortage—an environmental crisis strikes by surprise. This loss of resilience can be avoided by adaptive and flexible methods of resource management that take into account the often unpredictable variability of ecosystems.

Eighth Model: The Concept of Vulnerability

The concept of vulnerability provides an essential framework for understanding the consequences of environmental changes on human societies. It was first introduced by scientific inquiry into the causes of famines and, more recently, the impact of climate change. A classical study of environmental impact analyzes the consequences of a particular disturbance (a drought or a long-term change in the climate) for human societies. A study of vulnerability, by contrast, inverts the problem by analyzing the risk that a particular society will be adversely affected by a series of external shocks.

This risk largely depends on internal factors, which reduce or increase the ability of the society in question to respond or adapt to these disturbances. Vulnerability depends, then, on the disposition of individuals or groups to anticipate a disturbance, come to terms with it, resist it, and recover from it. This approach broadens the environmental analysis of the physical, biological, and chemical changes of a natural environment to incorporate economic, social, and political strategies of adaptation to these changes, as well as a society's capacity to restore its environment in the event it is degraded.

A simple analogy may make it easier to understand the importance of analyzing vulnerability. The acceptable ceiling of a loan does not depend solely on the absolute amount of the sum borrowed. It depends also on its value relative to the debtor's ability to repay the loan. This ability depends in turn on a series of characteristics of the debtor's situation: his present monthly income, the reliability of this source of revenue in the future, the value of his other assets, and the existence of a social network to which he can turn for assistance in time of need.

Similarly, the fact that a drought reduces the agricultural output of a nation or rural community by 10 or 15 percent does not by itself make it possible to predict a famine. The occurrence of a famine depends also on the existence of forms of social organization allowing households to find alternative sources of income and food, either by drawing upon reserve stocks, exchanging assets, or pursuing remunerative activities outside the agricultural sector.

A rise of fifteen inches in the level of the seas, accompanied by coastal flooding, is of little concern to a country like Switzerland, which is not exposed to this type of disturbance, or to the Netherlands, which anticipated the problem by raising its dikes. Bangladesh, by contrast, is extremely vulnerable: first, because it occupies a low-lying coastal alluvial plain; second, because farming, the principal source of income for its people, is liable to be very adversely affected by a massive incursion of salt water; and, finally, because its economy does not allow it either to rapidly mobilize sufficient capital to reconstruct infrastructure destroyed by a rise in the level of the oceans, or to make preventive investments in a system of coastal defenses. The vulnerability of a society depends therefore on its degree of exposure to risk, its awareness of potential problems, and its ability to anticipate and cope with such problems.

This ability depends in turn on a society's readiness in the face of a possible disturbance, its stock of available resources, and the variety of strategies for survival it has developed, as well as its degree of flexibility and adaptability. The more or less predictable character of an environmental problem also affects the vulnerability of a society. Natural hazards such as drought can rarely be foreseen, but a society that has put in place an effective network of assistance to disadvantaged populations and arranged for the prompt distribution of emergency supplies will be able to survive them. The general tendency of climate change at the present moment is plain enough, by contrast, but few steps are now being taken to reduce the vulnerability of certain economic sectors in the face of an increase in temperatures or in the frequency of extreme weather events.

Human societies are vulnerable to environmental surprises, such as a possible weakening of the Gulf Stream, the ocean current in the Atlantic that carries warm water from the Gulf of Mexico to northwestern Europe. Thanks to the evaporation of the waters of this warm current, winters are milder in the temperate regions of Europe than in regions on the same latitude in Canada. Thus Paris lies along the same latitude that passes through the middle of Newfoundland and the upper part of the Gaspé Peninsula, where winters are long and bitterly cold. The chance that climate change might lead to considerably colder winters in Europe is by no means impossible, since the like of it has already occurred, some ten thousand years ago. This time it would be due to a large inflow of fresh water from melting polar ice caps into the Atlantic. But since the probability of such an outcome is generally considered to be low, European countries are likely to be taken by surprise were it actually to occur.

An analysis of vulnerability must also take into account the various stressors that interact or operate simultaneously, and whose cumulative effect creates a certain risk. These often include external factors, which represent a particular threat, and internal factors, which represent the capacity of a society to respond to this threat. Some African communities, for example, are more vulnerable to famine than others because the social system has marginalized them to the point that they are deprived of access to economic and social resources that act as a buffer during a drought. A society may also be vulnerable to climatic changes if the agricultural and forest sectors of its economy are overly specialized in monocultures that are poorly adapted to higher (or, as in the case just discussed, lower) temperatures.

Reducing the vulnerability of a system requires that its resilience be increased, by improving its flexibility in mobilizing resources and by diversifying its function. It will also respond more effectively to disturbances the more freely and quickly information circulates, and the more rapidly this information is assimilated by decision makers. An early warning system for drought, for example, allows resources to be preventively mobilized, while responsible research on climatic changes makes it possible to develop plausible scenarios that permit an appropriate and timely political response. A society may therefore reduce its vulnerability by investing in environmental research, by being alert to changes in the information this research generates, and by acting upon this information in order to respond more promptly to both internal and external challenges.

The future of human societies depends not only on the manner in which they will be affected by changes in the environment (whether related to climate or something else), but also, and equally, on the severity of such changes and on the strategies societies adopt in order to mitigate them and adapt to them. The main elements of the analysis these challenges require us to undertake—nature, human demography, economy, technology, institutions, politics, social relations, and culture—have now been introduced. It remains for us to understand how they interact with one another, and to consider what tragedies are likely to unfold and what tragedies may yet be avoided.

I wished to understand the world of phenomena and physical forces in their complexity and mutual influence.
ALEXANDER VON HUMBOLDT

4 The Causes of Environmental Change

Reflection upon the causes of environmental change is probably as old as human civilization. In the fifth century B.C., Herodotus noted that the population of Lydia was too great in relation to what it could produce from the land, which had led to a famine lasting eighteen years. And in the first decades of the Christian era, Seneca the Younger, asserting the existence of a relationship between the population of Rome and the city's pollution, held that the growing number of kitchen stoves, the dust raised by traffic, and cremation of the dead on the outskirts of the city had made the air increasingly unfit to breathe.

It was not until the appearance of Thomas Malthus's *Essay on the Principle of Population* (1798), however, that a less anecdotal analysis of the relation between population and natural resources was attempted. The argument advanced by Malthus—that population growth is bound to outstrip the means of subsistence and therefore to lead to famines—had such an impact that still today the adjective "Malthusian" figures in virtually every discussion of birth control.

This argument was revived in the middle of the twentieth century, notably with the publication of *The Population Bomb* (1968) by the American biologist Paul Ehrlich, who likewise forecast an ecological crisis due to the growth of the world's population. Several years later, scenarios based on models of natural resource use were formulated by a group of scientists and thinkers known as the Club of Rome. These projections led them to predict, in *The Limits to Growth* (1972), a dark future for humanity in the event the world's population continued to grow rapidly.

In the view of some neo-Malthusians, responsibility for rapid demographic growth resides mainly with the world's poorest people. On a global scale, these are the developing countries; on a national scale, they are the marginal classes, whether disadvantaged ethnic groups or immigrant populations. This aspect of neo-Malthusian ideology has its roots in two cultural traditions: blaming the poor for their condition, due to excessively high fertility rates; and the bourgeois fear of being overrun by barbarian hordes massed at the city's gates.

The conviction that environmental degradation is linked to the demographic growth of the poor became so entrenched that for a long time research into other possible causes was not thought to be necessary. Neither the unbridled consumption of Western societies nor misguided policies that had led to mismanagement of natural resources were called into question. The solution to the problem was simple: exclude the fast-growing mass of marginal populations from lands rich in natural resources and replace their traditional livelihoods by modern economic activities, which, it was supposed, would be better able to sustainably exploit fragile ecosystems.

In the event, neither the dire predictions of Malthus nor those of other Cassandras since have come to pass. No one disputes that demographic growth increases the pressure on the environment. As the models presented in the previous chapter make clear, however, environmental changes have a far broader range of causes and involve a great many more mechanisms than Malthus suspected. A detailed analysis of the historical and geographic context in which such change has occurred unambiguously shows that it is always the result of a complex and distinctive combination of interacting factors. To be sure, demographic growth has considerably influenced the terrestrial environment over the course of both human history. But this growth is largely a consequence of other social and economic changes that have brought about a profound transformation of human societies, which has led in its turn to changes in the environment.

In seeking to illuminate the interlocking sources of environmental change, one needs to steer a middle course between simplistic approaches that pretend to identify a unique and universal cause (whether Judeo-Christian culture, capitalism, poverty, colonialism, overpopulation, or the patriarchal society) and objections that the task is irreducibly complex, in part because each situation is unique. The first sort of claim is misleading, and the second is not terribly useful. In the following sections I will try, first, to demonstrate the critical role of decision making in determining the nature and extent of human impact upon the environment, and then to identify the fundamental causes of the major types of environmental change.

Who Is Responsible?

Every environmental change is the result of decisions taken by agents, which is to say by persons or entities that have the power to take certain actions. These agents may be individuals or communities that manage their local environment (a household, for example, that decides to sort its trash for recycling); but they also include private enterprises (a chemical manufacturer that decides to invest in equipment for cleaning up atmospheric emissions from its factories, to transport its products by train or by truck, or to move its production facilities to a developing country where environmental standards are less strict) and public officials who conceive and put into effect policies that have an environmental impact at the local, regional, or global level. At the local level, planning agencies assign priorities to economic, social, and environmental objectives in evaluating development proposals. At the global level, international institutions negotiate multilateral environmental agreements concerning climate change, biodiversity, desertification, or fishing quotas, for example.

All these decisions have intentional or unintentional effects on the environment. A policy aimed at expanding a country's agricultural frontier and populating a territory covered by tropical forest deliberately attaches priority to the creation of new farmland rather than to the preservation of a natural ecosystem rich in biodiversity. This choice is founded on the conviction that the positive short-term effects of deforestation—an increase in food production and the extraction of forest resources, and the rise in living standards that follows from it—will outweigh its negative effects. As a rule, only those negative effects of a proposed policy that directly bear upon a country's own economy are taken into account; negative effects for countries elsewhere in the world are generally ignored.

By contrast, a policy of subsidizing agricultural production has the perverse and unintentional effect of favoring activities that may have harmful effects on the environment. Chemical pollution of the water and soil, though it is not the objective of such a policy, nonetheless is liable to affect a large population that is not compensated for its injuries, and therefore amounts to an externality.

Some environmental changes result from a decision that can clearly be attributed to an agent, such as the decision taken by President George W. Bush and his administration not to ratify the Kyoto Protocol on climate change. Most changes, however, are the result of a great many decisions taken by agents who do not confer with one another, who do not carry out the directives of a central leadership, who are motivated by diverse interests, and who adapt their behavior, sometimes in anticipation of such changes, to their perception of the context in which they are operating. Though each decision, in and of itself, has only a minor effect, and though the decisions of certain agents partly offset the harmful effects of the decisions of other agents, the aggregate effect of these decisions is to produce a change in the environment. Whereas some people throw trash out the windows of their car, others devote a part of their spare time to cleaning up the roadways. But when a great number of decisions converge, a change may rapidly occur without any particular actor being able to claim complete responsibility for it. This change is therefore an emergent property of the interactions between agents and their environment.

For example, the majority of the inhabitants of Western countries decide each morning to use their car to take their children to school or to go to work. While the environmental impact of each of these small trips is negligible, the cumulative effect of so many decisions on the scale of a city and its outlying areas is to substantially increase traffic congestion, air pollution, and fossil fuel consumption. When similar decisions are taken in a large number of cities and countries, the burning of this fossil fuel emits gases in quantities sufficient to contribute to a warming of the climate. In 1995, 777 million cars, trucks, and motorcycles were in use throughout the world. Each car traveled an average of 12,000 miles per year in the United States and between 7,000 and 9,000 in the European Union, emitting an average of 400 grams of carbon dioxide per mile traveled in the United States and 300 grams in the European Union. Worldwide, automobile transportation was responsible for atmospheric emissions totaling 1.7 billion tons of carbon dioxide in 1995.

The decisions taken by various public and private agents therefore have direct and indirect effects on the environment. Coming under the first category are decisions such as clearing a plot of forested land, dumping toxic waste into an unmonitored area, or harpooning a whale. Responsibility is readily assigned in this case because such decisions are taken by agents in immediate physical contact with the natural resource affected by their actions. Stopping at this level of responsibility misses the essential point, however. "Microdecisions," which directly affect the natural environment, are influenced by more fundamental decisions taken in the more or less recent past at higher levels of organization in a given society. These "macrodecisions" define a series of decision-making rules and programmatic procedures—associated with the functioning of markets and the legal system, industrial policies that more or less closely affect the environment, and the cultural and social heritage of a country—that shape interactions within society.

The peasant who cuts down a tree in the Amazonian forest, the driver who empties a dump truck filled with toxic waste, and the fisherman who shoots a whale with a harpoon are only soldiers carrying out the orders of their generals—orders that have become specific and precise during the course of their descent through the chain of command. In the environmental domain, these orders or signals, which prompt local agents to act, emanate from the economic, social, political, technological, scientific, and cultural spheres of society. At every level—global, regional, national, and local—these signals are enriched or impoverished, amplified or attenuated, and mixed with other signals, coherent or contradictory as the case may be. The agents at the end of the chain of command who wield the ax or harpoon, or who pull the lever, interpret the signals that reach them as a function of their own values, their perception of reality, their motivations, and their personal history.

The environmental impact of intensive agriculture in a European country, for example, is conditioned by the world price for cereals, regional subsidy policies, national levels of supply and demand, and local environmental policies that impose restrictions upon the use of liquid manure, the kind of farming practices promoted by local cooperatives, the size and type of individual farms, the agroecological characteristics of the land under cultivation, and so on.

In the majority of cases, an agent who operates at a local level in an ecosystem influences not at all, or only to a small degree, the decisions taken at a national, regional, or global level. Directives issuing from a higher level nonetheless determine the constraints and opportunities that influence local decisions. For example, the managers of a polluting factory have the power to choose between several production technologies whose environmental impact is more or less great. But they do not control all the relevant parameters of this decision, such as the cost of the various technologies available, the pollution standards set by national regulatory agencies, and price fluctuations on the world market, all of which will make this factory more or less competitive depending on the production technologies adopted by rival firms.

On the local scale and in the medium term, by contrast, individual agents have the power to modify certain parameters, by voting for a political representative who supports a program with which they are in agreement, for example, or by forming a local committee to sustainably manage a natural resource on which they depend, or by campaigning for alternative patterns of consumption and leisure activity.

Understanding the causes of environmental change requires going back to the factors that fundamentally influence the decisions taken by agents who have a direct impact on the environment. To resort once again to a military analogy, it is necessary not only to understand the impact of the tactical movements of individual soldiers, but also the way in which these movements reflect the strategies of the generals ordering them and the underlying motivations for waging the war. This is difficult to do, not only because of the large number of factors involved and the sometimes unpredictable interactions between them at different levels of social organization, but also because of variations between particular historical and geographical situations. The complexity of the problem is reduced somewhat, however, by virtue of the fact that in most cases a restricted series of interactions is responsible for a particular instance of environmental change.

In other words, the generals in the war over the environment rely repeatedly on a small number of tried-and-tested strategies that, while they are adapted in each case to a specific situation, nonetheless constitute a limited repertoire of maneuvers. As in chess, the movement of the attacking pieces is inspired by a few sequences of proven effectiveness.

The Five Syndromes

The various factors that lead to environmental change may be grouped into five great "syndromes," which is to say five recurrent patterns of association among a series of causes that interact to give rise to a certain impact on the environment. The first two are typical of developing countries, while the third and fourth occur more frequently in industrialized countries. The fifth one affects all of the world's countries.

Resource Scarcity and Overexploitation

The first syndrome occurs in weakened societies that have partly lost control of their destiny, and corresponds to situations in which a scarcity of natural resources increases the pressure on these resources in a vicious circle of environmental degradation. A familiar example is the local overexploitation of soils, forest resources, and minerals by a rural community in a poor country.

This syndrome is caused by rapid population growth (which increases the demand for natural resources), the appropriation of surplus production by a local elite or an external power group (which increases the pressure on resources), the application of technologies that are poorly suited to fragile ecosystems (the use of depth charges in fishing, intensive farming in mountainous areas), but also by a reduction in the size of the workforce (due to an aging population, emigration of the young, and so on) that makes certain kinds of maintenance or protection of natural resources impossible. In all these cases, the degradation of resources concentrates demand for those resources that remain, magnifying still further the stress placed on the environment.

The fundamental cause of this syndrome is generally unequal access to resources among users, combined with the intrinsic fragility of these resources. Environmental change therefore results as much from a dysfunction of the social system in distributing rights of access as from a scarcity of resources in absolute terms.

Loss of Resilience and Increased Vulnerability

The second syndrome appears when managers of natural resources—whether a nation, a business, a community, or an individual—experience an increase in vulnerability and a decrease in resilience. The entity in question therefore loses its ability to manage its environment in a sustainable manner and, in the case of external disturbance, finds itself caught up in a cycle of ecological degradation. This syndrome typically occurs in poor countries that undergo an episode of rapid environmental change. This is what happened in the Virunga National Park, north of Lake Kivu in the Democratic Republic of the Congo, which was dramatically degraded as a consequence of the settlement there of a million Rwandan refugees, forced to undertake a daily search for firewood.

The increase in vulnerability is generally associated with economic or social impoverishment, which may itself result from political marginalization. Economic factors such as an accumulation of debt, lack of credit facilities, and the absence of stable and diversified sources of income, as well as insufficient capital to cope with periods of crisis, make it impossible to attach priority to the preservation of natural resources. Social factors such as internal conflicts, the erosion of domestic alliances and associations that provide aid in times of need, or a substantial degree of dependence on external sources of assistance interfere with the ability to cope with economic or financial crisis, for example, or with extreme climatic conditions. Groups suffering from social or ethnic discrimination, or from epidemic disease, that then risk overexploiting their meager resources are also vulnerable, as well as societies battered by a succession of natural catastrophes (floods, earthquakes, and so on) or sociopolitical accidents (war, social tension, economic crisis) that weaken their resilience and force them to degrade their environment.

In this syndrome, the fundamental cause is therefore often to be sought in political events that have led to the economic and political exclusion of a social group, depriving it of any alternative to mismanaging its environmental resources.

Expansion and Intensification of Economic Activity

The third syndrome corresponds to situations of economic opportunity, which give rise to new activities or to an intensification of existing activities. Economic incentives encourage individuals to increase their consumption of goods and services, and private enterprises to increase production, despite the environmental consequences that flow from this. The societies most affected in this case are ones whose economies are expanding: the environmental dangers are known (though possibly underestimated), but they are perceived as negligible by comparison with the anticipated short-term welfare gains.

This syndrome is found mainly in industrialized countries and, more than any other, has affected the earth's environment in the twentieth century in the form of pollution of the air or water, climate change, destruction of the ozone layer, expansion of commercial agriculture at the expense of natural ecosystems, worldwide reduction of fish populations, and so on.

A variety of factors is responsible for the emergence of environmentally destructive economic opportunities: a lowering of the price of certain products that require natural resources for their manufacture; a fall in production costs, thanks to improved manufacturing methods or to more efficient technologies (without, however, any corresponding reduction of their environmental impact); investments in new sectors of activity that come into conflict with the preservation of natural resources (for example, the building of shopping malls and amusement parks on marshland or peat bogs that are rich in biodiversity); the construction of transportation infrastructure that provides access to fragile and formerly inaccessible ecosystems. The root cause of the problem is that market prices do not incorporate environmental costs, or do so only partially, particularly when the environmental impact affects public resources (the atmosphere or groundwater) rather than private resources (the land owned by an entrepreneur). Only new activities having weak environmental impact, or none at all, would seem attractive if prices reflected the full range of associated environmental costs.

For example, if the price of gas covered the cost of the atmospheric pollution caused by the use of cars and trucks, as well as the noise and congestion of traffic, trains would become the preferred means of transportation. In the United States, a gallon of gas costs less than a gallon of milk, with taxes accounting for scarcely more than a quarter of the price at the pump. Adding in the hidden costs of gas would have the effect of multiplying its price three, four, five, even six times. In Europe, tomatoes produced in the south of Spain are sold cheaply in Germany and the Netherlands, despite the expense of having to truck them almost a thousand miles to market. The people who live near the highways over which these tomatoes are transported may be said to indirectly subsidize them, by not demanding compensation for the environmental costs they incur in the form of pollution.

Demographic Shifts and Sociocultural Change

In the fourth syndrome, environmental changes are due to social, cultural, and institutional changes. Urbanization and the spread of modern Western culture have generated new and often contradictory relations with nature in much of the world. On the one hand, production enjoys economies of scale thanks to the concentration of a large mass of consumers in a small area, which makes the exploitation of natural resources more efficient and allows a higher value to be placed upon green spaces, with nature now becoming a source of new opportunities for recreation. On the other hand, urban garbage and waste pollute the atmosphere and blight the countryside, mass consumption grows without regard for its impact on natural resources, and metropolitan areas are degraded by the creation of a dense transport network and the relocation of polluting industries to the urban periphery.

Attempts to increase awareness of the need to protect the natural environment also come into conflict with the cult of individualism and a growing loss of connection with nature. A recent poll revealed that 90 percent of city dwellers in Great Britain are unaware that beer is made from barley, and 20 percent that yogurt is made from milk. Karl Marx long ago suggested that the capitalist mode of production reduces direct interactions between human beings and the natural environment from which they draw their sustenance, accelerating its degradation.

These social and cultural changes are accompanied by new institutional arrangements for regulating access to natural resources. Everywhere in the world, communal and artisanal management of nature has given way to private ownership of property and industrialization, along with the nationalization of certain natural resources and the disenfranchisement of those who traditionally used them.

The fundamental cause of this syndrome is a profound social and cultural transformation that separates mankind from nature and allows economic rationality to determine how natural resources are to be exploited. Industrial development has confined the natural environment to the role of being a supplier of resources, a repository for waste, a source of recreational space, and an opportunity for a novel sort of conservation (as though nature were now regarded as a vast al fresco museum where the genetic capital of animal and plant species can be preserved for future commercial use by industry).

Unintended Policy Consequences

The fifth syndrome of environmental change, and the one most frequently encountered, involves poorly conceived policies that, despite their praiseworthy social or economic objectives, turn out to have perverse effects on the environment. This syndrome operates at various levels of power—local, national, regional, and global—and often includes poor coordination between policies at different levels or between policies affecting different sectors of the economy.

A great number of policies having an adverse environmental impact are devised without the participation of the communities that manage the environment at the local level, and therefore do not always take into account the environmental knowledge these communities have developed. The overexploitation of oceans, forests, and water resources due to governmental subsidies, the drying up of the Aral Sea, the malign consequences of great dam-building projects throughout the world, and programs for colonizing the tropical forest in Brazil and Indonesia are all examples of this global phenomenon.

Among the most environmentally harmful policies are ones that rely on government subventions and tax incentives. Subventions are a way of assuring the survival and stability of strategically important sectors such as agriculture. If the objective is to guarantee a minimum level of food production or to sustainably manage farmland, subventions can be an appropriate policy tool; but if the primary purpose is to maximize agricultural productivity, they often have the effect of contaminating soils and aquifers with toxic substances, harming human health through the excessive use of pesticides, and exacerbating soil erosion while making it necessary each year to destroy crop surpluses.

Subventions may take the form of direct transfers to producers or consumers, manipulation of market prices (notably by means of tariffs), preferential fiscal policies (tax exemptions, credits, and deductions), a reduction in the cost of inputs (free access to natural resources), and complementary capital goods (such as highway infrastructure required by certain activities such as forestry and mining).

The eminent British environmental scientist Norman Myers has shown that many direct subsidies lead to a degradation of the environment, whether by encouraging overconsumption of water, overfishing, motor vehicle pollution, or the burning of fossil fuels. The ratio between government subsidies for renewable energy sources and subsidies for nonrenewable sources in the mid-1990s, Myers points out, was one to ten in the United States and in most member countries of the European Union. This imbalance gives an artificial economic advantage to the most polluting sources of energy and, by reducing the price of these sources, encourages neither the search for cheaper forms of energy nor the development of cleaner technologies.

It sometimes happens that the negative environmental effects of different subsidies accumulate within a particular sector of the economy. Agriculture in Europe and the United States is a good example: in extreme cases, multiple subsidies successively result in the clearing of new land for farming, an increase in productivity through the application of chemical fertilizers, irrigation using water priced at a tenth of its actual cost, the destruction of surplus crop yields, and the abandonment of previously cleared land, left to lie fallow; what is more, agricultural subsidies give developed countries an unfair advantage in competing with the agricultural exports of developing countries, which in turn helps to keep these parts of the world poor. Another telling example is provided by Canada, which in the 1980s subsidized the expansion and modernization of the fishing fleet in Newfoundland, only then to find itself obliged a few years later to pay fishermen to keep their boats at dock, the stocks of fish in coastal waters having been exhausted in the interval.

A sector in one country may benefit from indirect assistance when the environmental cost of its activity is borne, not by the sector itself, but by other countries that are not involved in this activity. Scandinavian countries, for example, indirectly subsidize the production of electricity in Great Britain since their forests suffer from the effects of the acid rain that results from it. Unlike environmental problems caused by direct subsidies, which are the product of governmental intervention in the market, in this case it is the absence of governmental intervention to force a polluting industry to assume financial responsibility for its external environmental costs that is the source of the problem.

On a global scale, the sectors most affected by perverse subsidies are, in decreasing order, highway transport, agriculture, fossil and nuclear energy production, public water supply, fishing, and forestry. Norman Myers has shown that every year, worldwide, some $2 trillion are spent on perverse subsidies, or roughly four times the annual monetary income of the 1.3 billion poorest people on the planet.

Poor governance of resources due to corruption, negligence, and incompetence on the part of policy makers who favor short-term private interests over the long-term public good also leads in many cases to predatory exploitation of the environment. It frequently happens, for example, that private groups are granted free access to common environmental property, whether for the purpose of extracting natural resources, exploiting the economic potential of fragile ecosystems, or discharging toxic products into them. Costly infrastructure is often constructed with public funds without regard for its disastrous environmental impact. In other cases, poor implementation of otherwise sound policies contributes to environmental change. Governmental instability and wars serving the narrow interests of an elite cause environmental conservation programs to be abandoned, bring in their train a flood of refugees who overexploit accessible resources in order to survive, and create a political climate that is unfavorable to the long-term management of natural resources.

All the syndromes we have just examined are characterized by a dysfunction in the conception and execution of policies, if only because of the inability of political policy makers to respond in a timely fashion to threats to the environment. The paradox is that regulatory agencies, when they function properly, are also the most efficient means for resolving many environmental problems.

The Interconnectedness of Syndromes

It is rare that a single syndrome or a single cause is responsible for a major environmental change. In the majority of cases, a change arises either from a series of successive events or from the action of several independent but simultaneous events. For example, the combination of a warming climate and the human modification of terrestrial vegetation significantly disturbs a great many biological species: global warming leads to a geographical shift of climate zones favorable to different animal and plant species, while agricultural and urban expansion fragments the habitat of these species, isolating populations from one another.

Quite often a combination of factors slowly modifies an ecosystem and prepares the way for a sudden and unpredictable event, which then rapidly brings about a radical modification of the environment. These factors gradually and invisibly erode an ecosystem's resilience, so that the appearance of a new factor is able abruptly to trigger a serious environmental problem. For example, decades of constructing asphalt parking lots and of pulling up hedges, together with farming practices that compact the land and leave it exposed in winter, have made the soil progressively more impermeable while increasing runoff. When a particularly rainy year next occurs, the damage caused by the floods that follow is substantial.

The famine of 1845–46 in Ireland illustrates how several independent factors—some of them gradual, others sudden—can cause a major catastrophe. The population of the country grew from 800,000 inhabitants in 1500 to 8.5 million in 1846. Unequal distribution of land had led to reliance upon the potato as the sole staple food for almost half of the population. Poverty and malnutrition, long widespread, had been growing worse since the middle of the eighteenth century.

In 1845, a sudden epidemic of mildew (a disease of potatoes and other plants) caused the loss of a part of the harvest that year, and the whole of the harvest the next year. The human impact of this catastrophe was exacerbated by the policies of the British government, which declined to intervene in the market for agricultural products. At the same time, in addition to refusing to subsidize a program of food assistance, it suspended investment in public works projects such as road construction, the only alternative sources of income for the starving peasantry, with a view to reducing the dependence of the poor on government assistance. Finally, a large share of Ireland's abundant cereal production continued to be exported to England, the native population not having sufficient income to buy the grain produced on its own territory.

About a million people died from hunger or from illnesses related to malnutrition. Another million Irish emigrated during those two years, along with 3 million more by the end of the nineteenth century, mainly to the United States. Other European countries struck by mildew in 1845 escaped famine on this scale, because their agricultural production was more diversified and the policies they adopted to meet the crisis were better suited to the situation.

In other cases, a sudden and unpredictable environmental crisis can cause a series of changes that render a society more resistant to future crises. Following the eruption in April 1815 of Tambora, a volcano in Indonesia, which sent enormous clouds of ash into the atmosphere, the year of 1816 was known throughout the world as "the year with no summer." New England, in the northeastern United States, was particularly affected, with snow in June and destructive freezes in July and August that produced famines. This singular event stimulated the development of new agricultural techniques and in many countries led to greater solidarity within communities.

Environmental changes are typically the consequence of interdependent, mutually reinforcing factors that act synergistically to modify an ecosystem. Their effect is therefore multiplicative rather than simply additive. The popularity of the automobile, an environmentally harmful vehicle, as a means of personal transportation in Western countries furnishes perhaps the foremost example of this phenomenon. In recent decades the value placed on individualism by Western culture has encouraged the development of time-saving devices. Owning a big car—a symbol of social status—has become more important in the popular mind than protecting the environment and promoting road safety and personal health (by riding bicycles, for example). Land-use policies have further strengthened individualistic values by creating a dispersed human habitat, while transportation policies have produced a low-density public transport network. Over time the car has therefore become increasingly attractive as a means of personal transportation, not least because both its price and the price of gas are indirectly subsidized by virtue of the fact that they do not incorporate the costs of their environmental impact. The high speeds made possible by an extensive highway network have made traveling by car more rapid than by public transport, notwithstanding that they are the major cause of traffic fatalities. By the end of the twentieth century, automobile accidents were responsible for the deaths of 1.2 million people in the world each year.

In combination with the freedom of movement that cars confer, individualistic values have stimulated consumer demand for more comfortable cars, cheaper gas, a more concentrated road network, more free parking, higher roadway speeds, and easy access to less densely populated rural and exurban areas. The mutual reinforcement of individualistic values and technology embodied by the personal automobile has produced such severe traffic congestion and, owing to urban pollution, such harmful consequences for public health that many people now find themselves the victims of this form of transportation. Many regions are now virtually paralyzed by automobile traffic, which traps some drivers in their vehicles for several hours each day. In London the average speed of motor vehicles was scarcely higher in 2000 than that of horse-drawn coaches a century earlier. There are signs, however, that public attitudes are now beginning to diverge from governmental policies. If demand for public transport continues to grow in the years ahead, a promising opportunity to reverse a long-standing environmental tendency will present itself.

This example illustrates an important aspect of environmental change and adds a level of complexity to its understanding. When the environmental impacts of certain human activities become significant, they lead to a modification of these activities by altering the demand for resources, the nature and variety of economic opportunities associated with the environment, the vulnerability of human societies, public attitudes toward the environment, resource management practices, and environmental policies. These feedbacks may be positive or negative, which is to say that they may amplify or weaken a particular environmental change. In the case of the automobile, these feedbacks for a long while were positive, then became negative once a threshold was exceeded. This phenomenon gives rise to a complex, nonlinear course of interaction between human activity and its environment that is difficult to predict.

The Interplay between Levels of Activity

A final source of complexity in the analysis of environmental change derives from the interaction of human activities at the local, regional, national, and global levels. These activities influence one another in both directions: from the global to the local, as the growing impact of international trade on individual communities testifies; and from the local to the global, since many international institutions and movements have their roots in local developments. For example, distinct episodes of environmental degradation in different parts of the planet were responsible for the rise of international ecological movements (Greenpeace and the World Wildlife Fund among them) and for the signing of international conventions on the environment. World Social Forums such as the ones in Porto Alegre, Brazil (in 2003), and Bombay (in 2004) bring attention to a range of local initiatives as well.

A typical sequence of events that has led to deforestation in the Amazon over the past forty years clearly illustrates the interaction between factors operating at several hierarchical levels. Until recently, the vast forests of the Amazon were occupied by indigenous peoples and colonists who collected the latex from a local tree, *Hevea brasiliensis*, used to make rubber (an activity that reached its peak between 1860 and 1912). These actors had little or no influence on the external factors that led to the colonization of the Amazon.

The decision by national governments, particularly in Brazil, to extend the agricultural frontier caused a massive migration of labor into the Amazon from the 1960s onward. These governments were motivated by a geopolitical desire to establish the influence of their respective states over the territory of the Amazon on the international level and by an interest in legitimating their own power at the national level. Their policies also satisfied certain economic imperatives, particularly the need to attract capital investment, to create new opportunities for national markets, and to pay off the nation's debt. Finally, they served to advance the purposes of the great landowners and industrialists, who were in a position to exploit the natural resources of the Amazon and who, owing to their close contacts with the government, were able to influence decisions affecting their interests.

From the interplay of these national factors emerged the regional development of the Amazon, based on the wealth extracted by the timber and mineral industries, the construction of large dams, government programs for resettling migrant workers along the Transamazonian highway and other new roads, and agricultural projects. All of this generated a spontaneous inflow of labor, supported by the development of basic infrastructure (secondary roads, electrification of rural areas, health services, provision of potable water). Between 1991 and 1994, 81 percent of deforestation in the Brazilian Amazon took place less than thirty miles from four great roads.

The migrant labor force came from the northeastern region of Brazil, the Nordeste, where landownership was unequally distributed and which had been hard-hit in the 1970s by a long drought, and from states to the south of the forest, where unstable prices had led large plantations to abandon the cultivation of coffee beans in favor of the mechanized and more profitable cultivation of soybeans, leaving many farmworkers landless and unemployed. A government slogan of the period in Brazil encouraged migration: "The Amazon—a land without men for men without land." Colonization of the Amazon was therefore a means of easing social tensions in the regions to the south of the Amazon and in the Nordeste.

At the local level, in the heart of the Amazon, an early phase marked mainly by timber harvesting and settlement of the land by small farmers, who were poor and incapable of sustaining high yields for more than a few years, was followed by the arrival of colonists having better access to capital. As beneficiaries of governmental subsidies and tax incentives, they invested in large ranches and devoted themselves to livestock farming. The high rate of inflation made the purchase and clearing of land in the Amazon a profitable speculative investment, putting these colonists into competition with small landowners and occasionally giving rise to armed conflict, highlighted by the assassination in 1988 of Chico Mendes, a humble rubber tapper who became known throughout the world for his passionate defense of the Amazonian forest. A few days before his murder, Mendes declared: "When I die I do not want flowers, for I know that you are going to take them from the forest. . . . Like me, the leaders of the rubber workers have worked to save the Amazonian forest and to demonstrate that progress without destruction is possible."

The winners in these disputes were able to expand their landholdings, while the losers were forced to move on and open up a new frontier deeper in the interior, where the land was less accessible and less expensive. Once again they proceeded to clear the forest. Behind the new frontier, small urban centers grew up, creating local markets for agricultural products, meat, and lumber. In recent years the harvesting of timber in the Amazon has increasingly been oriented toward the world market as stocks of valuable wood are exhausted in the forests of Southeast Asia.

Between 1960 and 1995, the joint action of these global, national, regional, and local factors led to the destruction of 220,000 square miles of forest in the Amazon, or 15 percent of the original forest. Deforestation continues at an even faster rate today in the Brazilian Amazon, stimulated by the growth of meat and soybean exports in 2002 and 2003. The economic benefits have been slight, however, since the output of the Brazilian Amazon as a whole represents only 4 percent of Brazil's gross national product and half of the region's 17 million inhabitants live in poverty.

A fortuitous combination of factors was required in order to produce an ecological disaster on this scale. If the world price of coffee had risen; or if a drought had not occurred in the Nordeste during the 1970s; or if the chainsaw, first introduced in 1917, had not been so widely used from the 1950s onward, permitting the rapid and profitable felling of large valuable trees—the Amazon would not have suffered the same fate.

Is Globalization to Blame?

The majority of environmental changes are caused by events on a global scale, among a great many others. While alterglobalist (and antiglobalist) movements claim that globalization is synonymous with ecological destruction, international economic institutions maintain, to the contrary, that globalization contributes to progress, and in particular to "ecological modernization." Who is right? The existence of environmental changes of worldwide scope does not by itself suffice to establish the culpability of globalization. We therefore need to examine the matter more closely.

The concept of globalization is too often reduced to its economic dimension, and especially to the liberalization of access to foreign markets. Globalization is more properly thought of as an entire set of processes that eliminate regional barriers to trade and migration, and that favor freer flows of people, capital, merchandise, and information throughout the world, as well as awareness of the fact that the world functions as a whole. It also includes the diffusion of new norms and values by the media, the emergence of a world civil society, the expansion of tourism, and a strengthening of international governance of the environment.

There is no doubt that the pace of globalization has accelerated in recent decades. Far from being a new phenomenon, however, it is the culmination of a series of events that began with the emergence of capitalism some four hundred years ago. Were not Marco Polo, who set out from Europe to the east in the thirteenth century, and after him Christopher Columbus, who set out to the west in the fifteenth century, the true pioneers of globalization? The Indian who saw Cortés disembark in Mexico in 1518, bringing with him massacres, diseases, and slavery—this Indian, whose gold was exported to Europe and whose ancestral traditions of agriculture were replaced by livestock breeding along the lines of contemporary practice in the Iberian peninsula—may well have been the first victim of globalization. Indeed, his way of life was much more profoundly disrupted by this primitive form of international trade (primitive in both senses of the term) than that of his present-day descendants, who are able to connect to the Internet from their homes in Mexico City and elsewhere. Globalization may therefore be said to have begun in the late fifteenth century and to have undergone two phases of accelerated growth, the first in the late nineteenth century and the second during the 1980s.

Owing to their multiform character, the effects of globalization do not always converge. Globalization has the power both to amplify and to weaken the causes of environmental change. The argument that globalization bears a large share of the responsibility for the planet's environmental deterioration runs as follows: markets are no longer local, but worldwide; world demand and supply regulate the exploitation of local resources that in many cases are essential for the proper functioning of ecosystems; whereas communities have always regulated the use of their land, today these decisions are dominated by the blind, anonymous, and soulless mechanisms of the international market. Thus, the tropical forests of Southeast Asia have been degraded to produce furniture made from local tree species valued by consumers in industrialized countries. The African rhinoceros finds itself in imminent danger of extinction because in East Asia its horn is supposed to have medicinal properties. Moreover, the liberalization of capital flows and the repeal of laws meant to protect local markets encourage investment in regions that enjoy competitive advantages. This leads in turn to an increased geographical specialization of economic activity, with the result that the diversity of the landscape is diminished, together with its ecological stability and aesthetic value.

But such specialization also favors the establishment of an appropriate balance between economic development and the ecological potential of a region, reducing the abusive exploitation of vulnerable ecosystems: the pressure on marginal and fragile lands diminishes as investment shifts in favor of areas with greater potential. Furthermore, globalization brings uniform standards of ecological performance, which allow consumers throughout the world to choose goods produced by methods whose sustainability has been certified. To be sure, these certification procedures need to be improved still further. But the continuing development of markets associated with the preservation of natural resources (the worldwide growth of ecotourism, for example, and private initiatives for the protection of endangered species) helps to create new economic opportunities and leads to greater efficiency in the use of natural resources.

Like urbanization, globalization widens the gap between consumers of natural resources and the land that produces these resources. Because an ecosystem's response to the pressures of production is not directly perceived by consumers, their behavior no longer adjusts itself to the possible degradation of ecosystems, and negative feedbacks that might otherwise be expected to moderate harmful exploitation of the environment are weakened. At the same time, the globalization of information allows people throughout the world to learn about environmentally destructive practices, wherever they occur, and to exert pressure for reform through nongovernmental and transnational organizations. Globalization also makes scientific knowledge and the technologies needed to correct ecological imbalances more widely available and more effective, for example, by improving the efficiency of energy use. Finally, international conventions make it possible to create a legal framework for the protection of the climate, biodiversity, and the planet's ozone layer.

Predicting the impact of globalization on a particular ecosystem is difficult because it may amplify or weaken the classical determinants of human impact (modes of consumption, production technologies, institutions, governance, population, values, and attitudes), depending on the case. It is precisely this uncertainty that arouses so much apprehension. But if communities lose a part of their autonomy, and so become more vulnerable to rises in the price of oil, currency devaluations, and financial crises as they are integrated in a global network, they profit at the same time from new values, new forms of expression, and new modes of production, thus diversifying the ways in which they can respond to external disruptions. It is easier for a community of poor farmers, for example, to take advantage of international support in an interconnected world.

All these things—the mixing of cultures, the dizzying increase in the number of possible paths of development, and the partial loss of control by individual communities over their destinies—are without doubt the true source of both the worries and the hopes that are felt today with regard to globalization.

Coping with global environmental change will necessarily involve coordinated actions on a worldwide scale, and therefore a strengthening of international institutions. But it is for just this reason that these institutions must actually be made to work for the good of all mankind, rather than to serve only the interests of a privileged few.

The Ultimate Cause of Environmental Change

The impact of human activity on the environment gained impetus in the twentieth century under the influence of a constellation of forces: unprecedented economic expansion, the reorganization of modes of industrial production, dramatic technological innovation, the rise of mass consumption on a grand scale, the development of new energy sources (in particular fossil fuels), rapid growth of the world's population, the geographical redistribution of this population through migration, rampant urbanization throughout the world, the increasing integration of national economies, international conflict, political and economic imperialism, the loss of a belief in the sacred character of nature, the development of a scientific view of the world, the dominance of a school of economic thought that for most of the twentieth century ignored nature, and a political ideology that treated economic growth as an imperative. All these factors, as the American environmental historian John McNeill has shown, reinforced one another to produce the present state of affairs.

But is it possible to go further than this, to trace the sources of environmental change to an ultimate cause? Because environmental history is the result of a multitude of decisions, as we have seen, it is at the level of the individual agents who make these decisions that a first cause must be sought. It seems reasonable to assume that individual decisions to consume a growing quantity of goods and services are responsible for a pattern of development that, in the course of a few centuries, has profoundly disturbed the terrestrial environment—even if, as with the problem of the chicken and the egg, the evolution of consumption has depended on all the factors just mentioned.

Beginning with the Industrial Revolution in Europe, the growth of consumption improved diet and permitted broader access to health care and hygienic products (soaps and washable cotton clothes, for example), which in turn increased life expectancy. The demographic transition to lower birthrates (in response to lower rates of mortality) took several decades to be completed, however, with the result that population grew during this period. The increase in per capita consumption and in the number of people required a corresponding increase in production, while resource scarcity stimulated major technological innovations. The resources essential to economic expansion were sought in ever more distant lands, which aroused imperialist and colonial ambitions, at great cost to certain indigenous peoples who found themselves cast into a deepening spiral of poverty, vulnerability, and environmental degradation. The insistence upon more intensive production and consumption prompted rural populations to move to cities, favoring the emergence of secular values that regard nature as an instrument of human needs and purposes.

In the space of only a few hundred years, the rate of increase in global consumption overtook that of population. According to John McNeill's reckoning, from 1890 to 1990 the earth's population human grew by four times while the consumption of industrial products grew by forty times, energy consumption by sixteen, water consumption by nine, the consumption of fish by thirty-five, and the total volume of the world economy by fourteen. This discrepancy between population growth and the growth of consumption is still more pronounced in economically advanced countries, which are responsible for a large share of global environmental change, whereas consumption in the world's poorest countries scarcely keeps pace with demographic growth.

The appropriation of ever more natural resources to meet growing consumption needs is not only the experience of Western civilization in recent centuries; it is also, on a smaller scale, the experience of many ancient societies in different places and at different times. It does not depend fundamentally on modes of economic or political organization, since the predatory use of natural resources, without regard for the ecological destruction that accompanies it, has occurred under both capitalist and socialist systems. All of the world's great civilizations, from the Far East to Mesoamerica, have adversely modified their natural environment to one degree or another. The abusive exploitation of natural resources is therefore not uniquely the legacy of the Judeo-Christian tradition, even if this tradition has greatly surpassed all others in its environmental impact. The history of all civilizations, ancient and modern, strongly suggests that an insatiable, though not unalterable, human desire to possess material goods lies at the root of the worsening degradation of the natural world.

The Roots of Consumption

Researchers have estimated that, at the end of the twentieth century, daily energy consumption in a "primitive" tribe (the !Kung bushmen of the Kalahari, a desert region in southwest Africa) whose members were in good health was on the order of 0.11 kilowatts per person as opposed to 11 kilowatts per person in the United States. The consumption of various items (not including water) was about 8 pounds per person per day in this tribe, and about 130 pounds per person in the United States. Ninety-nine percent of the consumption by the !Kung bushmen was composed of recyclable products, as against only 0.6 percent in the United States.

The American psychologist Abraham Maslow distinguished several types of fundamental needs that motivate consumption and that can be ranked in a hierarchy. The first three are, in ascending order, physiological (the need to be fed, sheltered, and so on), psychological (the need to assure one's physical and psychological security, one's identity), and social (the need to belong to a group, to love and to be loved). Above these are two more, associated with self-esteem (the need to establish one's personal dignity, to feel self-confidence, to be respected, to have social standing) and self-actualization (the need to realize one's abilities, to grow as a person, to give meaning to the events of one's life). In Maslow's hierarchy, "higher" needs acquire an important role only once "lower" needs are satisfied: needs evolve as a function of the development of individuals and of societies, passing from an overriding concern with survival to personal aims associated with the style and quality of one's life.

The American economist Lawrence Abbott, on the other hand, opposed "generic" to "derived" needs. Abbott defined generic needs as ones that are intrinsic to human nature and insatiable. Derived needs, by contrast, are momentary technological responses to generic needs that are capable of being satisfied. Thus, for example, the generic need of each person to be able to move about at will, and without impediment, could be perfectly fulfilled only if human beings were to acquire the gift of ubiquity; but the possession of an efficient and comfortable car (a derived need) allows it to be satisfied in the currently most advanced possible way. Perhaps in a few decades the personal helicopter (a technologically more sophisticated form of the same derived need) will become the next way in which the human need to move around rapidly will temporarily be satisfied.

A third classification of needs is due to the English economist John Maynard Keynes, who distinguished between "absolute" needs, which are felt regardless of one's situation (the need to eat, to quench one's thirst, and so on), and "relative" needs, whose satisfaction, as Keynes put it, "lifts us above, makes us feel superior to our fellows" (the need, for example, to wear fashionable clothes or to have a youthful appearance and an athletic physique). Whereas absolute needs can be satisfied, relative needs are insatiable: the gap between the present level of consumption and the level of consumption to which one aspires can never be closed, for these two levels are continually rising in parallel with one another.

In industrialized countries, the urge to satisfy absolute needs associated with subsistence, good health, a decent standard of life, and a dignified, free, and creative life accounts for only a fraction of household consumption. Indeed, the consumption of certain goods reaches proportions that threaten the health of individuals: in the United States, two-thirds of the population is overweight, and a third obese. The medical risks posed by obesity and excessive food consumption have created in their turn a new form of consumption: whereas each year the American food industry spends $30 billion on advertising to persuade consumers to eat still more, these same consumers now spend $33 billion a year in a mostly futile attempt to lose weight. One kind of unhealthy consumption, which has a measurable impact on the environment (in the form of pressure on food production, the use of nonbiodegradable packaging, and so on), is therefore supplemented by a second kind of consumption aimed at solving the problem created by the first, which also has a measurable impact on the environment (the manufacture of medicines to lose weight or to treat health problems associated with obesity, waste incineration and disposal, and so on).

Historically, increases in consumption are associated with the settlement of human societies. Prehistoric societies were organized in small-scale nomadic clans, each consisting of a set of extended families whose relations to one another favored an egalitarian social structure. These societies subsisted on hunting and the gathering of the fruits of nature. Both the mobility and the social organization of these component clans limited consumption: frequent displacements in response to the migration cycles of game, the seasonal rhythms of plant growth, and the fluctuating availability of water acted as brake on the accumulation of individual property, given the rudimentary means of transportation in prehistoric times.

The domestication of plants and the invention of agriculture, together with a profound modification of the social and political organization of ancient nomadic societies, gave birth to the first urban settlements. Among the many changes that accompanied the appearance of these more complex societies, the differentiation of social status and the emergence of a leadership class seem to have had a decisive impact on consumption and the environment.

Consumption and Social Competition

Along with the new hierarchical structure of settled societies, there appeared the need to establish, make known, and consolidate one's social status, power, and privileges through the consumption of goods that conferred prestige. Ruling elites encouraged an increase in production, the surplus from which was devoted to acquiring or creating the new goods that enlarged their power and their capital, while reducing their vulnerability in the face of external risks: in periods of political troubles or climatic fluctuations, such goods—most of which, by contrast with food reserves, did not deteriorate over time—could be exchanged for food. This development, which was to be the point of departure for a chronic pattern of excessive production in the future history of human societies, was made possible by a recognition of the abstract concept of "value," associated with things that were highly desired, notably power and privilege.

Consumption and the will to power are therefore intimately related. Surely it is not accidental that the head of a tribe or family has always claimed for himself the right to eat before others and more than others; that the chief executive officer of a large company pays himself a considerably higher salary than that of his employees; that the most powerful country on the planet both consumes and pollutes more than any other. The social history of mankind confirms the view that the consumption of relative goods in Keynes's sense is in large part determined by competition over social status. In societies of mass consumption, this process has been considerably amplified since the end of World War II, with the result that the majority of people in the wealthy countries of the world are now able to acquire both nonessential and prestige goods.

The social and symbolic stimulation of consumption has had a pernicious effect, for it opens the way to unlimited escalation: the consumption of a prestige good by a person seeking to improve his social standing is imitated by people who feel that they are entitled to similar status, immediately provoking still higher consumption on the part of the person who originally aspired to a higher standing, and so on. In other words, as the French economist Alain Cotta puts it, "someone else's luxury becomes your necessity." In a sense, consumption has replaced the chivalrous contests of the medieval world as the preferred form of social competition.

A few decades ago, the consumer one measured oneself against was a neighbor—a bit wealthier than oneself, to be sure, but still from the same social class. With the disappearance of traditional neighborhoods, as well as the entry of women into business, models of consumption were transposed to the workplace, where people found themselves exposed to a broader spectrum of social classes than they had known in their neighborhoods. The American sociologist Juliet Schor points out that well-known actors and the wealthy characters they play in television series have become models for the consumption habits of the people who watch them. In the United States today, there is a demonstrable correlation between the number of hours spent watching television and the amount of money spent shopping.

When people of moderate income model their shopping preferences on those of a few affluent persons, real or imaginary, the gap between the level of consumption to which they aspire and the income of which they actually dispose widens to the point that it can never be closed, creating a permanent sense of dissatisfaction. This dialectic of relative needs, and the chronic frustration it implies, makes it impossible for wants to be fulfilled. The endless growth of individual consumption that follows from this situation is the tragedy of modern consumer society, which finds its ecological expression in Ghandi's formula: "Nature can satisfy all of our needs but none of our greeds."

Advertising strategies readily exploit these impulses of competition and emulation. The mode of consumption both reflects and accentuates social inequalities. In modern society, the right sort of automobile is felt to be the ideal way for men to cut a handsome and dashing figure. A few decades ago, fur coats fulfilled this function for well-to-do women. It is the same with space shuttles for nations, cows for Masai herdsmen, fashionable clothes for adolescents, and expensive toys for children.

Every culture and subculture perpetually redefines the styles of consumption associated with different levels of social prestige. In all cultures, and throughout human history, what the American economic sociologist Thorstein Veblen called conspicuous consumption has been the favored way of exhibiting proofs of status. The media, with their dependence on commercial advertising, jointly serve to promote the emulation of certain forms of prestige consumption on a global scale. Certain goods that once were considered to be inessential luxuries—the television, cellular telephone, personal computer, and so on—appear to many people today as necessities.

Other motivations encourage the belief that personal well-being may be increased through greater consumption as well: the search for comfort, which reduces personal tensions through the satisfaction of elementary needs; the search for stimulation, which combats boredom by means of novelty, change, risk, and surprise; the search for short-term pleasures and new experiences, which respond to a range of needs that have been subjectively created over the course of a person's life or under the influence of advertising. Yet recent research unambiguously shows that a culture of materialism has deleterious consequences for emotional well-being.

Changing patterns of social organization also stimulate consumption. A reduction in the number of persons living under the same roof, as family members become more dispersed geographically, leads to an increase in the consumption of housing units, household appliances, and goods circulating within the home (books and newspapers, for example). The same number of people, redistributed over a greater number of separate residences, inevitably consumes more than it did before.

Redirecting Consumption

The environmental impact of consumption depends not only on its quantity but also on its composition, which is to say the amount of energy and resources that goes into making consumer goods and the waste and pollution emissions that they create. It seems unrealistic to try to drastically reduce consumption in quantitative terms, for the human hunger for social status, comfort, pleasure, and a sense of achievement will continue to cause relative needs—whose satisfaction provides us with a feeling of superiority in relation to others—to grow. A more realistic ambition would be to try to reorient consumption toward goods having a smaller environmental impact.

This means redirecting human needs toward a type of consumption that is rich in qualitative rather than quantitative terms (so that getting a good education, for example, matters more than owning a big car) and whose consequences for the environment are comparatively benign (visiting a museum or going for a run, rather than watching sports car or motorcycle racing events). Respect for others, social status, and self-fulfillment would then be achieved through a simple and altruistic way of life rather than by a selfish obsession with superficial luxuries. Since most needs are cultural in origin, they can in principle be reformed by modifying their cultural and social context. The most useful instruments for bringing about such a change are therefore the institutions responsible for transmitting cultural values to future generations, in particular the family, the educational system, and the media.

National energy policies can offer powerful incentives for rechanneling consumption as well. For example, the price of gas and, to a lesser degree, electricity are lower in the United States than in Japan: whereas in the 1990s per capita income in the United States was only 15 percent higher than in Japan, the energy consumption of an American household was twice that of a Japanese household. The relatively high cost of electricity has encouraged Japanese industry to use energy more efficiently and Japanese consumers to adopt a more economical style of living. Whereas in America 85 percent of all travel was by car, in Japan this figure was 55 percent.

Differences of this type reflect the variety of cultural preferences and ways of life that may be formed over time in response to government policies and information conveyed by the media.

> . . . Mankind faces a crossroads. One path leads to despair and utter hopelessness. The other, to total extinction. Let us pray we have the wisdom to choose correctly.
>
> **WOODY ALLEN**

5 Ecological Degradation or Restoration

While some complex societies and civilizations—Egypt, China, Japan, and Java among them—have given proofs of impressive longevity, in others the ecological basis of the economy became degraded to the point that they collapsed and disappeared. Degradation on this scale has occurred in small Polynesian island societies of the South Pacific (notably Easter Island) as well as in the civilizations of the Anasazi Indians of the American Southwest, the Maya in Central America, and possibly Angkor in Southeast Asia. The disappearance of the Viking colonies in Greenland in the Middle Ages is explained in part by overgrazing, but also by a failure to adapt to the cooling climate of the period.

Many societies have disappeared after having colonized a formerly uninhabited island where the fauna had not yet learned to mistrust human predators. These animals were rapidly exterminated by the new settlers, who found a readily accessible supply of nourishment on their arrival. The absence of a regulatory mechanism that would have made it possible to use natural resources in a sustainable way led to the demise of these societies, which include the various Polynesian peoples that settled the Mangaia, Henderson, Necker, Pitcairn, and Norfolk islands in the South Pacific, as well as the Hawaiian island of Kahoʻolawe and the south island of New Zealand.

Other societies collapsed on account of their vulnerability to an arid climate, in which the native vegetation only slowly recovers after its destruction by humans and the exposed soils rapidly erode. These conditions were fatal to the Amerindian civilization of the Anasazi (or Pueblo culture), in what is now the state of New Mexico.

Notwithstanding the differences between individual cases, most of these societies, whether long enduring or not, followed a similar course of development whose main features at the outset included a complex and stratified social structure, centralized political power, a growing population and territory subject to a common authority, a variety of specialized and coordinated economic activities, intensive agriculture, urbanization, a developed commercial network, small wars fought at the boundaries of the territory, and domination by a religious and military elite. This pattern of development led all these societies to a critical moment when a choice had to be made between creating new institutions and technologies or pursuing a policy of growth without any corresponding adjustment of their economic, social, and political systems, whose environmental basis therefore risked severe degradation.

In some cases, the importance of this moment of decision was not perceived: key actors lacked the will or the capacity to make appropriate choices, or else external events (climatic disturbances, for example) pushed them onto a path of ecological destruction. In the extreme case, the economic and political decline these societies experienced brought about their collapse; in other cases, appropriate choices were made at the right moment, inaugurating a phase of innovation and ecological restoration. These societies were able to go forward on a sustainable basis and subsequently enjoyed a certain prosperity.

There is no doubt that such junctures presented themselves many times in the history of ancient societies; that in many cases there were second chances and possibilities for recovery; and that the precise nature of the crisis differed from society to society depending on the historical period of development in which it occurred. The image of a juncture, or fork in the road, that is reached at a particular moment is a simplification, of course, because the multitude of decisions that guide a society toward one path of development rather than another are spread out over a relatively long period of time. Even so, it helps us think more clearly about the factors that tilt certain societies toward environmental degradation—a prelude to the weakening, perhaps even the collapse of their civilization—and others toward sustainable development. Why does a society follow one path rather than the other? Which events trigger a process of institutional and technological innovation and which ones inhibit such a process?

Before answering these questions, it will be useful to examine two historical episodes that illustrate the extreme cases bracketing the range of possible environmental trajectories: the collapse of Maya civilization in the ninth century and the reforestation in Europe in the nineteenth century.

The Collapse of Maya Civilization

Over the last several centuries, colonizers, adventurers, and archaeologists have discovered to their amazement the silent ruins of ancient Maya kingdoms at Copán, Tik'al, Kalak'mul, and Chich'én Itzá in the heart of the tropical forest of Guatemala, Honduras, Belize, and the Yucatán in Mexico. These vestiges of a great ancient civilization in Central America, and its sudden disappearance for reasons that were long mysterious, caused many traditional assumptions about the New World to be revised.

Archaeological research since the mid-nineteenth century has revealed that Maya civilization was the most sophisticated of all the ancient civilizations of the Americas, attaining a high level of development in the domains of art, science, and social and political organization. The Maya acquired a detailed knowledge of astronomy, devised several types of calendar, invented a complete system of writing, constructed monumental edifices, and created a network of administrative and ceremonial centers covering a vast territory. They also produced a substantial written literature, most of which was burned in 1562 by an overzealous Franciscan missionary.

The plains formerly occupied by the Maya in Central America stretch from the Chiapas Mountains in Guatemala to the Gulf of Mexico and the Caribbean Sea. The annual rainfall varies from more than 150 inches in the southern plain, in Belize and the northern parts of Guatemala, to scarcely more than 15 inches on the semi-arid coast of the northwest Yucatán Peninsula. At the end of the eighth century, between 3 and 10 million people (depending on the estimate) lived in this territory, with densities as high as 775 persons per square mile in certain areas, but at the time of the Spanish conquest there were no more than a few hundred thousand, the result of a rapid demographic collapse that took place seven centuries earlier.

The first settlements of sedentary farmers in the region go back to 2500

B.C., but the clearing of the land began in earnest only a thousand years later. The first signs of a complex form of social and political organization appeared between 650 and 400 B.C. There followed a period of spectacular changes, between 400 B.C. and 250 A.D., during which the population grew rapidly, commerce developed, a distinctive architectural style emerged, and religious and royal symbols appeared along with the first evidence of writing. Traces of military conflict dating from this period can still be seen today.

The first important modifications of the landscape, mainly the result of draining fields and other hydraulic projects, date from this period. Classic Maya civilization was well established by 300 A.D. and continued to develop, finally attaining a phase of maturity between 600 and 800. At the zenith of the development of Maya civilization, three-quarters of the territory was cleared for agriculture. Between forty and fifty autonomous, yet interacting, centers sharing a common culture had their own dynastic lines. Kings were supposed to possess virtually supernatural powers. The centers were the seat of royal courts and the site of ceremonial functions, which consisted in part in the conspicuous and lavish consumption of prestige luxury goods, some of them brought from far away.

The monumental architecture of the Maya, the spectacular ruins of which can be admired still today, was meant to assert the supremacy of the royal court, both within centers and in competition with neighboring courts. The peasants, who represented 80 to 90 percent of the population, were dispersed over a radius of a few dozen miles around the central temple complex. Even though the Maya world never enjoyed any real political unity, its separate kingdoms constituted a large cultural system that surely deserves to be ranked among the great ancient civilizations of the world. From 550 onward, however, these kingdoms formed a series of shifting alliances and proceeded to tear one other apart in an interminable series of wars, particularly during the seventh and eighth centuries.

Then, abruptly, in the ninth century, Maya civilization experienced social, political, and demographic collapse. This sudden decline was manifested in several ways: kings, nobles, and dynastic traditions disappeared; the exchange of goods and information among neighboring centers dwindled; the construction of new temples, funerary monuments, and steles was interrupted; production ceased of prestige goods for the elite, such as pottery, sculpted stones, and jade jewelry; the use of writing systems and calendars from the Classic Period came to an end, and the administrative class withered away; and population in the centers and the countryside rapidly dropped (most of the great centers were abandoned after 800; the population fell by 80 percent in a century and by 90 to 99 percent shortly thereafter). In certain kingdoms, these events occurred very rapidly. In others, the collapse of elites and, with it, the splendor of their rule were sudden, but the decline of the peasantry was more gradual, continuing in some cases for 150 years after the disappearance of the royal court.

After its eclipse, Maya society never managed to reestablish itself in the southern and western parts of its initial territories: ruins, and lands that previously had been intensively cultivated, were now covered over by tropical forest. The inability of Maya society to recover by making the necessary adjustments to its social and economic system—in other words, its lack of resilience—seems in retrospect as surprising as its rapid collapse.

By contrast, in the northwestern part of Maya territory, and to the east along the coast of the Yucatán Peninsula, a few new centers were established after the collapse of the ninth century, as we know from the accounts of the first Spanish invaders at the beginning of the sixteenth century. Before the arrival of Cortés, these small kingdoms had themselves experienced successive collapses, followed in each case by the partial migration of the peasant population to another center. The last small independent Maya kingdom, numbering no more than a tens of thousands, survived until the end of the seventeenth century in the heart of the Guatemalan forest; it was crushed by the Spanish army in 1697 after a few hours of fighting. These residual centers were smaller and more isolated than the ones that existed during the Classic Period. Maya society had lost virtually all of its former magnificence, strength, and size.

The Causes of the Collapse of Maya Civilization

Many explanations have been proposed to account for the collapse of Maya civilization. They include endogenous causes (erosion of the soil, overpopulation, unsuitable agricultural practices, peasant revolt, military conflict, religious beliefs and superstitions) and exogenous causes (disturbance of commercial links with the outside world, invasion by neighboring peoples from Mexico, devastating earthquakes and hurricanes, climate change, epidemics, and invasion by insects detrimental to crops).

The meticulous reconstruction of Maya civilization by several generations of archaeologists has revealed that several events were jointly responsible for its collapse. Although areas of uncertainty and disagreement persist, it has nonetheless been possible to piece together a coherent picture of what happened. The American archaeologist David Webster has persuasively argued that four factors—some of them interacting with one another—triggered the crisis: environmental degradation caused by unsuitable agricultural practices; the destabilizing effect of internal sociopolitical tensions, local wars, and acute competition over resources; popular rejection of royal ideology and institutions caused by the inability of these institutions to provide effective solutions to the problems of the age; and a series of droughts, particularly severe in the ninth and tenth centuries.

Moreover, these factors created or exacerbated secondary disturbances, such as difficulties in assuring a reliable supply of water, social unrest, and diseases associated with malnutrition. The relative weight of these causes, and the circumstances and chronology of the collapse of the various kingdoms, varied slightly from region to region.

Over Maya territory as a whole, population growth led to rapid deforestation and the cultivation of marginal land whose soils were poor and fragile, the comparatively rich land of the valleys having already been occupied. The population continued to grow, making it necessary to abandon the traditional system of itinerant slash-and-burn agriculture, by which the land was left to lie fallow for long periods to permit the natural restoration of the soil's fertility, in favor of almost permanent cultivation. With the resulting loss of fertility came a decline in agricultural yields: the absence of large domestic animals made it impossible to use animal manure as organic fertilizer, and the soil, nowhere in this tropical region very rich to begin with, was exposed to a significant degree of erosion once stripped of its forest cover.

In several places—though not at Tik'al, the largest Maya center, and to only a very small extent at Copán, another major center—hillsides were terraced to combat soil erosion. Swamps were drained and filled in to enlarge the amount of arable land. But these costly, labor-intensive measures were insufficient, and came too late, to meet the dietary needs of a constantly growing population and to prevent the environment from being severely degraded. Study of ancient sediments in the lakes near Maya centers reveals evidence of rapid erosion and impoverishment of soil nutrients that coincided with the deforestation of the Maya territory. This pattern of destruction reached its height in about 800. Deforestation also led to a scarcity of timber for construction and wood for cooking, as well as to the disappearance of game, the Maya's sole source of animal protein.

The demographic saturation of agricultural land was intensified by the concentration of farming populations around the leading religious and royal centers. In these places an area extending at least one or two days' walk from the central temple complex was continuously settled, with densities of more than 1,300 persons per square mile in the neighborhood of Tik'al and about 2,400 per square mile around Copán.

This high level of concentration was due in part to a desire on the part of the ruling elite to increase the number of farmers from whom a surplus could be extracted in order to supply the royal court, and whose labor could be used for the construction of temples and for fighting wars: in the absence of an efficient system of transport, this population had to be settled in the immediate periphery of the centers. While the nobles and kings, being well-fed, were in good health, tall, and long-lived (some of them reaching seventy years of age), the undernourished peasants were small in stature and seldom lived beyond the age of thirty-five.

The recurrent conflicts among kingdoms also favored the concentration of the population, which sought security and protection near the centers, leaving unoccupied areas that might otherwise have served as buffer zones against enemy raids. The decline in agricultural productivity and the environmental crisis caused by overexploitation of the land probably aggravated conflicts both within kingdoms and among neighboring kingdoms. Social tensions were not slow in appearing between a densely settled and underfed peasant class desperately working poor farmland and privileged ruling classes that reserved the best land for themselves. Competition for scarce resources—particularly land, labor, and political power—must also have led to wars between kingdoms, as the many military conflicts of which we have evidence in the seventh and eighth centuries, just prior to the collapse, suggest. The lack of a regionally integrated political system stood in the way of a peaceful settlement of these often destructive and costly conflicts. Wars not only permitted territorial conquest; they also enabled kings to reaffirm their social status and to justify their privileges—in the event of victory, at least, failing which they were sacrificed in ceremonies organized by their conquerors. Success in warfare therefore formed an important part of royal ideology.

The collapse of Maya societies may be attributed as much to the inability of these societies to promote technological innovation in order to cope with demographic pressures as to population growth itself. Many solutions to the agricultural crisis were possible. Certain options were not explored as fully as they might have been, such as switching to crops that were more productive and made fewer demands on the relatively weak fertility of the soil. Manioc and sweet potatoes, for example, were grown in small quantities, while corn remained the favored crop. The expansion of farmland through the construction of terraces and the draining of wetlands was intermittent and, so far as we can tell, ineffective. Despite their impressive intellectual achievements, the Maya seem to have shown little aptitude for developing advanced technologies. The wheel, metal tools, and animal traction were unknown to them. And while there is no question that the Maya were competent farmers, in the absence of major technological innovation the short-term adaptive strategies they pursued were bound to be inadequate to solve their long-term problems.

The comparatively inefficient system of agriculture practiced by the Maya did not allow them to produce substantial surpluses. This situation stands in sharp contrast with that of ancient Egypt, where use of the plow, animal traction, and irrigation, as well as the elaborate social organization that more sophisticated kinds of agriculture require (including centralized distribution of rights of access to land and water), permitted each farmer to produce as much as five to six times what was necessary for his family's survival. The low agricultural surplus characteristic of Maya societies exposed food production to severe shortfalls from the effects of droughts, hurricanes, insect pests, and social and political conflict.

Agricultural vulnerability was exacerbated by the absence of domestic animals (other than turkeys, ducks, and dogs), which elsewhere in the world constituted reserve supplies of food and a source of organic fertilizer. By an irony of history, the native peoples of the Americas were handicapped in this regard by the fact that earlier, around 11,000 B.C., their ancestors may have contributed (in combination with climate change) to the massive extinction of large mammals through hunting. Animal species that might have been domesticated had all but disappeared. Making matters worse, the lack of an effective system of transport prohibited agricultural trade with neighboring peoples, and in any case the hot and humid climate of this tropical region made it difficult to preserve harvests from one year to the next. What small agricultural surplus there was seems to have been extracted in the form of taxes and offerings by political and religious elites in order to provide for royal families, administrators, warriors, and workers who were engaged in the building of temples and palaces.

Although the upper classes had managed to free themselves from the burdens of day-to-day survival, as the complex structure and splendor of Maya society plainly show, they nonetheless failed to adapt their behavior to long-term environmental constraints. If the labor and resources devoted to the construction of temples had been used instead to construct an efficient and sustainable system of terraced agriculture, and if elites had developed techniques for restoring the fertility of the soil, rather than engaging in extravagant competitions for the purpose of enhancing their own prestige, Maya society might have survived. But ruling dynasties and other institutions failed to take steps in response to the emerging agricultural and environmental crisis. Crucially, no centralized system of agricultural management was put in place when the first signs of soil erosion appeared. The only intervention of the Maya leadership—beyond taking a share of agricultural production from the starving peasants—assumed the form of ritual invocations of supernatural forces, human sacrifices, symbolic ball games, and other religious ceremonies.

Furthermore, it seems probable that the court's evident inability to devise pragmatic solutions and safeguard the natural order produced social unrest.

Afflicted by a high mortality rate, low life expectancy, malnutrition, disease, and a decline in soil fertility, the peasants appear to have placed the blame for their ills on the kings and priests, who, having claimed for themselves supernatural powers, ended up being the first victims of internal tensions aroused by the degradation of the environment and subsequently amplified by warfare.

Crowning this sequence of unhappy events was a long period of severe drought between 800 and 1000—the longest and most intense drought in several thousand years, with a few especially dry intervals of between three and nine years. Because the Maya depended on rain and surface water for potable supplies, this climatic episode would have particularly affected communities in the driest regions of the Yucatán, in the north. But it was the population centers in the south, in the wettest areas covered by tropical forests, that collapsed first; the kingdoms in the north did not experience significant depopulation until at least one hundred years later. The reason for this apparent paradox seems to be that aquifers were less accessible in the mountainous regions of the south, where the relatively high elevation of the land was an insuperable obstacle to digging wells, and the porous karstic soil conserved few natural reservoirs of surface water. Here we see once again that civilizations are vulnerable to the conjunction of simultaneous events, some of them independent of one another, whose combined effect is apt to be fatal, above all when they are unforeseeable.

One of the most striking aspects of the collapse of Maya civilization is that its extraordinary architectural, political, and intellectual achievements had attained their peak only a few decades earlier. The enormous pyramids, sumptuous palaces, and remarkable engravings and writings were only a brilliant facade that masked profound structural weaknesses—critical defects that an inattentive visitor would surely have failed to notice and that perhaps were not altogether obvious to the Maya rulers themselves. The end of Maya civilization was therefore inscribed in the most visible signs of its success. For an observer concerned only with the cultural, intellectual, and military power of this civilization, the suggestion that collapse was imminent would have seemed utterly improbable.

A thousand years later, in another part of the world, a quite different story was to unfold, though it, too, began with demographic growth and deforestation on a large scale.

Reforestation in Europe in the Nineteenth Century

Forests covered almost half of the earth's land ten thousand years ago, but they cover less than a third of it today. In the early nineteenth century, however, significant reforestation began to occur in certain countries of Europe, through both natural regeneration and replanting. Much the same phenomenon was subsequently observed in the United States, first in New England and then in the rest of the country. More recently, vast reforestation programs have been launched in China. This pattern of events is consistent with the environmental Kuznets curve. The fact that it is ongoing, and appears even in some developing countries, suggests that under certain conditions an ominous environmental tendency can be reversed and that economic and demographic expansion need not necessarily be accompanied by a diminution of natural capital.

The forest transition that took place in a number of European countries in the nineteenth and twentieth centuries—notably Denmark, Switzerland, Belgium, and France—is attributable to a set of essentially technological and institutional factors: the development of new agricultural and forestry techniques, which improved yields and made it possible to produce more on a smaller territory; a reduction in the need for wood by industry thanks to substitute sources of energy, particularly coal and, later, oil; a gradual adjustment of land use to environmental constraints made possible by improved transportation, with the result that agriculture and livestock rearing became concentrated on the most fertile land and marginal land was abandoned; enforcement of government regulations mandating the protection of natural and forest resources, in response to a perceived environmental crisis; a rise in urban population as a consequence of industrialization, accompanied by a corresponding depopulation of rural regions; political pressure brought by private citizens, for whom the quality of life and of the environment had come to assume greater value than in the past; and geopolitical events, independent of the local exploitation of natural resources, that served to increase national cohesion and strengthen the popular will to manage a country's natural capital in a rational way.

Whereas an earlier wave of reforestation in Europe, in the fourteenth century, had coincided with a general demographic collapse caused by the Black Death, reforestation in the nineteenth century was accompanied by rapid population growth. The factors I have just mentioned conceal significant variation between countries and periods, however, that has been studied by the British geographer Alexander Mather and his colleagues. In Denmark, for example, the area covered by forest is now three times greater than it was two hundred years ago. In the eighteenth century, Denmark suffered an ecological crisis that was widely recognized. By 1800 only 4 percent of the original forest cover remained (the Napoleonic Wars subsequently disrupted the importation of firewood from the duchy of Holstein in the south to Copenhagen). The resulting shortages led to a doubling of the price of wood between 1780 and 1800, triggered a rapid response by the government, and created a favorable climate for radical reform of natural resource management. The Forest Preservation Act (1805) privatized public lands on the condition that the new owners protected and maintained them, while also planting trees on land that had been cleared. Livestock were barred from both royal and private forests. By way of compensation, peasants who had lost grazing rights for their livestock in newly privatized forests were authorized to clear common woodlands where the tree cover was thin.

Scientific techniques of forest management, first devised during the Enlightenment, were borrowed from Germany beginning in 1763 and widely disseminated by forestry schools and university programs of forest management. The cause of reforestation did not really gain impetus until 1864, however, with the military annexation by Prussia and Austria of Schleswig-Holstein, one of the most heavily wooded regions of Denmark. This defeat aroused widespread patriotic support for a plan to reaffirm the country's independence by optimizing the exploitation of natural resources. Thus, for example, heaths in the province of Jutland were converted into farmland and forest by local associations, aided by state subsidies that by 1901 had grown to cover 33 percent of the costs of reforestation. Because the additional income earned by farmers for planting trees, in addition to tending their crops, was reinvested in modern equipment and farming methods, reforestation in Denmark was accompanied by a strong increase in agricultural production.

An analogous situation can be observed in Switzerland, where the forest cover has almost doubled since the mid-nineteenth century. During the years 1830–50, floods had caused great damage. The Swiss Forestry Society succeeded in convincing political authorities that these inundations were attributable at least in part to the damage that forests had sustained in the course of the preceding century. A report published in 1862 concluded that deforestation in the Alps was responsible for the more irregular rate of flow of the country's rivers, increasing the risk of avalanches and rock slides. New floods in 1868, which claimed fifty victims, seemed to confirm this analysis.

In response to a gathering ecological crisis, the federal government launched programs for reforesting and for limiting forest use in alpine zones. The Forestry Police Act (1876) required farmers to obtain permission to cut down trees and then immediately to replant either the area that had been cleared or an equivalent area nearby. Furthermore, the state modified the traditional rights of forest use, requiring farmers to reforest bare soils, especially on steep slopes, even in the case of land that was privately owned.

This legislation was supported by an official statement regarding the role played by deforestation in bringing about the ecological crisis that was indispensable in winning acceptance by both local leaders and the public for the new rules of forest management, whose adoption was further encouraged by the passage of constitutional amendments aimed at modernizing the federal system and by technological advances in agriculture. The establishment of the Helvetic Confederation in 1848 led to the creation of a government department of forestry, a university-level school of forest management, and a formal inquiry into the causes of the latest round of floods. In all these ways, the new federal state signaled its determination to actively participate in the management of natural resources.

At the same time, marginal agricultural land was abandoned on the slopes of the Alps as a consequence of the decline in the farming population and, though to a lesser degree than in Denmark, the rural exodus associated with industrialization. The displacement of livestock rearing from mountain pastures to valleys permitted a natural regeneration of the forests, along with the growing use of coal for fuel in place of wood, thanks in large part to the development of railroads, which made coal easier to import.

A similar sequence of events took place in Belgium, in the Ardennes. There deforestation had accelerated during the seventeenth and eighteenth centuries in order to satisfy the fuel needs of local foundries and tanneries, as well as of industries in the valleys of the Meuse and the Sambre. Medical progress had enabled the population to grow rapidly. Two events triggered a forest transition: Belgium's declaration of independence in 1830, which created a desire to take charge of the management of the country's natural resources, and a profound agricultural crisis that culminated in the severe famine of 1847–48, the joint result of a shortage of wood for cooking, diseases affecting rye and the potato (contemporary with the famine in Ireland), insufficient food production, and a decline in small family businesses that manufactured linen goods.

Vigorous intervention by the new Belgian state obliged villages to sell heaths and other tracts of open and uncultivated land that could be farmed or reforested to private buyers. By 1910 the area occupied by these old communal grazing lands had been reduced by 90 percent. Other measures aimed at supporting agriculture and forest management were taken, such as the promotion of fertilizer (lime in particular), the construction of roads to serve the countryside, the annulment of laws regulating trade in fertilizers and cereals, the distribution of Scots pine to rural communities, and so on.

Reforestation did not really get under way, however, until agricultural production in more fertile regions had sharply increased and the government repealed various measures intended to support farmers in marginal regions. A change in laws affecting landownership and the discovery of new coal deposits, which provided an alternative fuel source to wood, also promoted reforestation. The appearance of American wheat on the European market and the strong demand for labor in the textile and steel industries at the end of the nineteenth century gave rise to a sizable rural exodus as well. Abandoned farmland in the Ardennes was replanted. And the development of a rail transportation network allowed the region, which had long specialized in livestock rearing, to export milk and butter and to import staple foods.

In France, to take a final example, the forest cover is almost twice today what it was in 1830. Forests there have now regained the area lost since the fourteenth century, though their composition and geographic distribution are quite different. The pattern of forest transition in France was very similar to the one we have seen in neighboring countries. In this case, the ecological crisis was associated mainly with the erosion, sometimes catastrophic, of soils in the early nineteenth century, particularly in Champagne and Lorraine, the Causses region of the Massif Central, Provence, and the Alps. A new forestry policy was subsequently adopted whose main principles were laid down in the *Code forestier* of 1827.

Unlike what happened in other countries, however, the direct intervention of the French state in the management of forests, formerly the responsibility of local villages and conducted in accordance with long-standing custom, ran into vigorous opposition from the peasantry. The underlying purpose of the *Code forestier*—to arrest the degradation of forests, particularly in marginal mountainous zones—came into conflict with the immediate needs of rural communities for fuel and farmland where population pressures were great (in the Pyrenees, the Alps, and the Jura). In practice, the *Code* encouraged the harvesting of timber at the expense of other forest products. The struggle to restore forests also set urban and industrial elites against the rural population in a battle over control of natural resources. The state therefore had to resort to coercive measures. Resistance to the new form of forest management gradually died out as demographic decline took hold in the countryside.

In France, as elsewhere, the forest transition was made possible by a geographic reorganization of agricultural production, on both the national and the local scale, that resulted in a better balance between farming activities and the land's agroecological potential. Grazing in some mountain areas was abandoned in favor of irrigated agriculture in the valleys. The extensive herding of goats and sheep in areas of natural vegetation was replaced by the intensive rearing of bovine animals in meadowlands. Agriculture, having been abandoned in regions with the poorest soil, where only the subsistence needs of the local population had previously justified its presence, became concentrated and intensified in the most fertile regions. The development of mechanized transportation and a market economy enabled the inhabitants of these marginal regions to specialize in various manufactures while allowing the forest to reclaim a part of the land.

This policy of forest management in France was not the first: in 1669 Colbert had issued an *Ordonnance forestière* consisting of more than five hundred articles. But this ordinance was only partially put into effect in the outlying regions of the country—which were also the most heavily wooded—for the socioeconomic and political circumstances of the period were not favorable to a thoroughgoing reform of natural resource management. Governmental intervention is therefore a necessary, though not a sufficient, condition for the success of a forest transition.

Ecological Crisis Followed by Restoration

The European experience shows that environmental degradation can be checked, indeed reversed, and that ecological restoration can go hand in hand with economic and demographic expansion. In each of the foregoing cases, three major steps occur. Initially, the perception of a severe ecological crisis triggers a reaction, arousing leaders to intervene and to prepare the public to accept reforms, sometimes with the aid of a certain amount of dramatization. Next, the enactment of new legislation profoundly modifies the ways in which natural resources are managed. It then becomes possible, in a third stage, to implement these reforms without incurring socially unacceptable costs, thanks to technological innovation in the agricultural, forestry, and transportation sectors, and to the creation of new opportunities for industry. This process of modernization is accompanied by both social changes (urbanization and the abandonment of marginal regions) and political changes (countries take charge of their own destiny, developing scientific institutions for the management of resources). Several of these elements were sorely lacking in the case of the Maya, where degradation of the natural environment was allowed to continue, ultimately with fatal consequences.

It is nonetheless disturbing to observe, in all these cases, that the upheavals and high social and ecological costs that follow from an environmental crisis appear to be necessary (though not sufficient) to persuade societies to mount an adequate response. Must a profound crisis always be undergone in order to move from a course of harmful development to a sustainable one? The fact that the concept of "creative destruction" (due to the Austrian economist Joseph Schumpeter, who noted the incessant replacement of firms, products, and methods of production by more efficient technologies and forms of industrial organization) can be applied to many environmental problems, where destructive exploitation of natural resources leads to a crisis that then serves as a stimulus for innovations in resource management, may seem to suggest that this sequence of events is indeed unavoidable.

The risk nonetheless remains that an environmental crisis will have irreversible consequences (for example, a reduction in biodiversity) or that a political response will come too late, so that high costs are impossible to escape. It is also the case that many natural systems give evidence of great inertia: certain human actions, though they no longer occur, continue to have effects for decades, even centuries, afterward. Even so, there is an important lesson to be drawn from the environmental history of Europe in the nineteenth century. Everywhere that a period of reforestation followed one of deforestation, it was at least in part the result of vigorous governmental intervention, coordinated at all levels. In no part of the world has the renewal of forestland occurred by itself.

In the United States, President Theodore Roosevelt was horrified to discover that American forests were being cut down at a rate five times greater than their rate of regeneration. In order to avoid the catastrophic economic consequences of a foreseeable exhaustion of wood resources, Roosevelt created the U.S. Forest Service in 1905 and won approval from Congress for legislation that obliged landowners to protect their forests.

In China, catastrophic flooding from the Chang Jiang (Yangtze River) in 1998, which forced hundreds of millions of peasants to flee their villages, persuaded the government to restrict the exploitation of forests and to replant entire slopes of hills and mountains in order to limit runoff from torrential rains. (This policy has had the unintended result of pushing the Chinese timber industry into neighboring Burma, which now has one of the highest rates of deforestation in the world.)

In Thailand and the Philippines, landslides and devastating floods caused by the denuding of mountain soils led the governments of these countries in the early 1990s to prohibit certain forms of timber exploitation. Logging bans have been fairly successful and, although some illegal clearing continues, rates of deforestation have gone down significantly.

Restoration of the natural environment has never been the spontaneous result of economic growth. Nor is there any reason to suppose that farmland or pastures created today at the expense of tropical forest will one day be abandoned and reforested: the deforested land along the Mediterranean, for example, has never regained its tree cover. Moreover, new forests (for example, ones dominated by a single species of conifer) are poorer in biodiversity than their predecessors. Deforestation often causes irreparable damage to fauna, flora, and soils.

Thirteen Factors Influencing a Transition to Sustainability

Whatever the historical context or the geographical scale of a particular environmental situation, thirteen factors influence the course followed by a society as it approaches a moment of choice between a path of sustainable development (Europe in the nineteenth century, for example) and one of ecological degradation (the Maya of the Classic Period). The irony is that the number thirteen is thought to bring bad luck in European civilization, whereas it was considered a favorable number among the Maya.

These factors have to do with the quality of available information about the state of the environment, the motivation of actors to manage their resources in a sustainable manner, and the ability of these actors to put into effect a rational and long-term policy for managing the natural environment:

1. *The characteristics of the natural environment.* Certain regions of the world suffer from intrinsic ecological constraints—arid climate, poor and erodible soil, a scarcity of water resources, an abundance of parasites or other conditions favorable to the spread of disease, and so on. Not only is the productivity of natural resources weak in these regions; natural capital regenerates slowly there as well.
2. *Recognition of the early warning signs of an ecological crisis by resource managers.* If indications of impending environmental change fail to be detected, owing to the absence of an effective system for monitoring environmental conditions, an imminent ecological crisis becomes an invisible one as well. Recognizing such signs is more difficult not only with regard to complex ecosystems and to resources displaying migratory behavior (whales, for example), but also when interannual variability of environmental conditions—notably in the case of the climate in certain regions of the world—masks evidence of degradation.
3. *The communication of information about the environment to policy makers.* The transmission of urgent warning signals must be rapid and effective. In complex urban societies, the growing distance between policy makers and agents in direct contact with natural resources means that information about early signs of the exhaustion of a resource or the disturbance of an ecosystem sometimes reaches policy makers only after a long delay. During this time such information may also be altered, simplified, or distorted, either because of a lack of relevant expertise on the part of intermediaries or because they have a vested interest in preventing government agencies from interfering with local resource management.
4. *The diagnosis of the causes of a particular environmental problem.* Once the problem has been recognized, diagnosis of its causes—and therefore of

the action needing to be taken—must be correct. This often obliges policy makers to oppose conventional wisdom. The Maya probably saw that their land was being impoverished, but because they attributed the problem to religious causes, they sought a solution in sacrifices and offerings to supernatural forces. Even in less extreme cases, successfully identifying the causes of complex natural processes may be difficult.

5. *The will of policy makers to intervene*. Where the extent of environmental degradation has been accurately gauged and the reasons for it correctly diagnosed, policy makers may yet fail to recognize the importance of the problem, even deny its very existence. This occurs when a ruling elite defends short-term interests that diverge from those of a population affected by the threat to the environment. Thus, for example, a commitment to protecting the interests of the American oil industry is probably responsible for President George W. Bush's refusal to carry out national and international policies for reducing climate changes set in motion by human activity.
6. *The capacity for technological innovation*. An adequate response to environmental crisis often includes technological solutions. The Maya were scarcely acquainted with advanced technologies. In Europe in the nineteenth century, by contrast, the intensification of agriculture in the most fertile regions, the promotion of new industries, and the development of a system of mechanized transport all contributed to the success of reforestation policies.
7. *The diffusion of external innovations*. Not all innovations are the product of local ingenuity. The borrowing of ideas and solutions from neighboring societies has been a common practice throughout human history. The domestication of plants for agriculture, for example, originated independently in five regions of the world and subsequently spread to every other region. The speed with which new ideas about ways to diminish human impact on the environment are diffused depends on the state of networks of communication and the frequency of contact between populations having different levels of technological sophistication.
8. *The power and flexibility of institutions*. Technological innovation is often only part of the solution, however, and needs to be supplemented by fresh policies and the institutional reform of resource management practices. Successfully putting these new policies into effect requires that the state move rapidly to devise and enforce new regulations throughout its domain.
9. *The existence of a production surplus*. Innovation assumes prior investment in research and development, and the ability to absorb the risks that are inherent in it. This requires in turn a certain level of surplus production, and therefore a robust economy.
10. *Economic integration across regions characterized by a diversity of resources*. Some degree of economic or political cooperation between different regions

is necessary if the physical security needed for innovation—as well as access to a variety of resources, ecosystems, and social and human capital—is to be achieved. In France in the nineteenth century, regional specialization was an essential factor in the success of reforestation. By contrast, incessant warfare among Maya kingdoms dispersed resources and diverted attention from the search for solutions to their common environmental problems.

11. *The rate of environmental change*. For a society to be able to modify its socioeconomic structure, environmental change must be gradual: technological and institutional innovation takes time. In the modern history of Western countries, major technological innovations have succeeded one another over a period of several decades. The inevitable delay in adopting new technologies is a significant constraint, looking to the future, particularly with regard to the diffusion of new techniques of energy production, whether in the domain of transportation (hydrogen-powered cars) or the production of domestic energy (from wind and solar sources). Societies are more vulnerable to environmental changes requiring a rapid response than to gradual changes.
12. *Stability during the period of transition*. When catastrophic events unrelated to environmental degradation occur during the period of transition to another mode of economic development (natural catastrophes, such as a prolonged drought; political events, such as social unrest or war), the chance combination of a gradual process of environmental alteration and a sudden external disturbance may drag the society down into a vicious spiral of ecological degradation and socioeconomic decline.
13. *The presence of influential public figures*. In all the countries of Europe where forest transition was successful, to name only one example, influential and charismatic public figures played a crucial role in the conception, promotion, and implementation of new methods for managing forests.

For a transition to sustainable development to be successfully made in a timely manner, these thirteen conditions must be satisfied. The implementation of the Montreal Protocol, signed in 1987, which provides a mechanism for checking the destruction of the ozone layer, is an important recent demonstration that this is possible. By the same token, however, failure in meeting just one of these conditions is enough to substantially reduce the likelihood that a transition to sound resource management practices will occur, or at least makes such a transition more difficult. Some factors are more readily controlled than others, of course. The outcome of a movement in the direction of ecological restoration and sustainable development is therefore not something that can be settled by the behavior of a few privileged actors. It is a matter for society as a whole to decide and act upon.

Forests precede people, deserts follow them.
FRANÇOIS-RENÉ DE CHATEAUBRIAND

6 The Nature of Environmental Change

An old Indian fable relates that six men, all of them blind, one day encountered an elephant. The first touched the animal's flank: "It is an animal," he declared. The second seized its trunk and concluded that it was a giant serpent. The third felt the point of the tusk: "This animal is as dangerous as a saber," he said. The fourth man, after having embraced one of the animal's legs, thought that it was a tree. The fifth, feeling the pachyderm's enormous ear, declared that it was a fan, or perhaps a flying carpet. The sixth, after having caught hold of the animal's tail, was convinced that the elephant was actually an old rope. They then began to quarrel among themselves over the nature of this strange beast. Awakened by their cries, a rajah came and said: "How can each of you be certain that you are right? The elephant is a large animal, and each of you has touched only a part of its body. If you put the parts together, you may be able to arrive at the truth."

Several lessons can be drawn from this fable: human beings have trouble understanding complex realities that they cannot directly observe; they nonetheless arrive at conclusions, which are often erroneous; several points of view are therefore necessary in order to approach the truth.

Despite the patient accumulation of observations made using ever more sophisticated instruments, many uncertainties persist concerning the nature and scope of certain environmental changes. It is sometimes difficult to determine whether changes observed over a short period of time signal the onset of a tendency caused by human activity or whether they belong instead to a natural cycle that introduces a certain variability into environmental conditions. These uncertainties are apt to be used as a pretext for not implementing costly policy measures against an apparent instance of environmental degradation.

When the Desert Advances . . .

The debate over whether desertification has occurred in Sahelian Africa in the last half-century offers an instructive example of the difficulty that is often encountered in distinguishing a chronic and debilitating tendency from natural cycles of variability within an ecosystem. It also shows, in dramatic fashion, how an inadequate understanding of a presumed environmental problem can sometimes lead to misguided policy interventions that subsequently become part of the problem.

The Sahara is rich in archaeological vestiges that attest to the ancient human occupation of this region, now a desert. The frescoes of Tassili n'Ajjer in Algeria prove, among other things, that fauna typically found in savannas occupied the Sahara a few thousand years ago. The banks of ancient freshwater lakes, dry today, are strewn with pottery and other artifacts. A humid period lasting about thirty-five hundred years, from roughly 7000 to 3500 B.C., when the Sahara was green, was followed by a dry period that persists to the present day—a cycle of climatic variations that has been repeated many times, for tens of thousands of years. This cycle has a natural astronomical cause and so does not count as an instance of desertification due to human activity.

After an exceptionally wet period in the 1950s and 1960s, the Sahel—a transitional zone between desert and savanna on the southern edge of the Sahara—has undergone a series of severe droughts over the past thirty-five years, from 1970 to 1974 and then 1976 to 1993. An intrinsic feature of the semi-arid climate of this region is its strong variability from one year to the next (some years recording as little as 40 percent of the average annual rainfall) and from one area to another.

Desertification is defined as a degradation of land in arid and semi-arid zones, over several decades, that results from the interaction of human activity and climatic variability. Degradation in this case consists in a diminution or enduring loss of biological and economic productivity due to an inappropriate use of land for agriculture, grazing, and forestry. It is associated with the erosion of soils by wind and water; deterioration of the physical, chemical, and biological properties of the soils; and a long-term loss of plant cover.

The annual rainfall of the Sahel varies from four inches per year in the steppes in the southern part of the Sahara to thirty inches per year in the woody savannas farther south. These rains are concentrated in a period lasting from two to four months a year. The traditional way of life rests on a coexistence between transhumant pastoralism and a cereal agriculture based on millet and sorghum, both well adapted to dry conditions, and, along the Niger River, on rice. A whole range of land uses is found in this region, from nomadic herdsman to settled farmer, with various degrees of mobility and integration between livestock rearing and cereal cultivation. Over time the socioeconomic systems of the Sahel have developed a substantial degree of resilience in the face of climatic extremes, even though each drought brings its share of destruction and suffering.

Reality or Myth?

In 1907, a forestry mission dispatched by the French colonial administration of Upper Senegal and Niger concluded that the Sahara was creeping southward as a result of human activity—a conclusion subsequently endorsed by many colonial reports and travel accounts. In 1935, a report to the Royal Geographical Society in London by E. P. Stebbing, a professor of forestry at Edinburgh University, on his recent trip to West Africa confirmed the Sahara's advance, a phenomenon that he attributed to destructive agricultural practices and demographic growth.

Local officials reacted to Stebbing's claim with skepticism, not least because his observations had been made during the dry season, when travelers and foreign experts found it easier to get around: to visitors who had never observed the explosion of greenery that followed the first rains, the landscape appeared to be a wasteland. Then came two decades of exceptional rainfall (1950–67), and the question of desertification was forgotten.

With the great droughts of the 1970s and 1980s, however, it was forcefully revived. A British ecologist, Hugh Lamprey, traveled to Sudan in 1975 as part of a United Nations survey charged with assessing the extent of the desert's encroachment. Lamprey compared the landscape that year, which followed a severe drought, with the plant cover depicted on a map of Sudan's vegetation in 1958, in the midst of an unusually wet period. On this basis, he concluded that the Sahara had advanced between 55 and 60 miles to the south, or an average of 3.4 miles per year. This figure was adopted by governments, international aid and development agencies, and the media, and even made its way into textbooks before being revised upwards to 5.6 miles per year by George H. W. Bush, then vice president of the United States, in a March 1986 speech, and finally to almost 11 miles per year by Barber Conable, the president of the World Bank, in a 1989 speech. The original figure that Lamprey had come up with was based on the study of a part of Sudan, not of the Sahel as a whole; on observations involving two years, each characterized by exceptional climatic conditions; and on the unavoidably imprecise comparison of a rudimentary map with aerial observations.

Until the early 1990s there were few alternative estimates, most of these (based on questionnaires filled out by government officials in the countries concerned and a compilation of the opinions of 250 experts in an atlas of desertification) notable for their lack of methodological rigor. As early as 1988, however, a World Bank report (apparently unknown to its president) concluded that the extent of desertification had been exaggerated; that there was little proof that semi-arid regions were affected by desertification; and that there were few technologies available to deal with the situation, which amounted to claiming that nevertheless there was a problem.

The question of desertification attracted the attention of other international institutions as well, first with the creation of the United Nations Environmental Programme (UNEP) in 1972, one of whose principal mandates was the battle against desertification. Next came the United Nations Conference on Desertification (UNCOD) held in Nairobi in 1977, followed by agreement to include an article on the subject in Agenda 21, the declaration on environment and development issued by the Earth Summit held in Rio de Janeiro in 1992, and finally the United Nations Convention to Combat Desertification (UNCCD), which came into effect in December 1996. These discussions relied on similarly doubtful statistics: more than a third of the planet, it was claimed, was threatened by desertification, with a billion people in danger of suffering from its effects.

Today it is clear that the cause of desertification gained international momentum more on account of the humanitarian tragedies caused by the droughts of the 1970s in Africa than by any real advance of the desert established by precise measurements. Already in 1991, in the journal *Science*, researchers at the National Aeronautics and Space Administration (NASA) in the United States had published an analysis of data from observation satellites that showed the southern boundary of the Sahara advancing and retreating from one year to the next in response to irregular fluctuations of rainfall, without the least evidence of a constant trend. The maximum extent of these fluctuations over the course of a decade, between 1980 and 1990, was 155 miles of latitude, and the areas affected by drought became productive again once the rains regained their normal level—proof of the ecological resilience of semi-arid zones in the face of drought.

It was then discovered that a small part of the Sahara's expansion could not be explained by variations in rainfall. Was it due to the cumulative impact of a series of droughts that had eroded the resilience of the vegetation? Or was it the result of human impact? Before an answer could be found, however, the tendency to expansion vanished during the course of the following decade. Indeed, by the end of the 1990s, observation satellite data seemed to show that the vegetation cover in the Sahel as a whole had slightly increased.

But there was a catch: the satellites used in these regional studies collected data over spatial units ranging from six-tenths of a mile to five miles—too large a scale to distinguish all the details of plant cover and soil, which meant the possibility could not be excluded that analysis at finer spatial resolutions would disclose evidence of vegetal degradation or of soil erosion by wind and rain. It was therefore necessary to reconstruct the recent history of vegetation in various parts of the Sahel using more precise data.

One researcher analyzed aerial views of Sudan taken between 1943 and 1994, either from airplanes or, beginning in the 1960s, from American spy satellites. This information made it possible to distinguish individual trees in the landscape and to study their evolution over the previous fifty years. No diminution of the plant cover was found, even in the case of the sites studied by Lamprey.

Another researcher traveled on foot more than eleven hundred miles through the southern part of Senegal, visiting 135 villages in the course of a year. In each village he inventoried the trees and interviewed two persons over the age of sixty-five who had resided in the village all their lives. He drew upon their recollections to reconstruct the tree cover at the time of the great drought that lasted from 1942 to 1949, which inevitably had left a mark on their memory, and concluded that the density and diversity of the cover had significantly diminished in this part of Senegal, though more because of climate change than of human activity. In this area of the Sahel, plant species typical of savannas were found to have retreated southward at an average rate of 550 to 650 yards per year, or ten times less quickly than Lamprey had estimated in Sudan.

Aerial photographs from the 1950s, mostly taken by the National Geographic Institute of Senegal (then still a French colony), made it possible to reconstruct the plant cover during this decade in several regions of the Sahel. A comparison with the present situation led to divergent conclusions: while certain regions had experienced a net loss of natural vegetation, other neighboring regions saw a net gain where drought had degraded vegetation in the interval. In zones that had been stripped of their plant cover during the great drought of the 1970s, for example, dunes formerly stabilized by this vegetation were set adrift and then re-anchored by the growth of a new plant cover.

Still more surprising, agricultural yields in Burkina Faso have continually increased since 1960, particularly for cereals such as millet and sorghum that are grown on unirrigated fields with little or no chemical fertilizer. The productivity of the soil therefore did not diminish everywhere in the Sahel, even if some amount of erosion has occurred in various places, either naturally or as a consequence of human activity. On balance, despite signs in some places of local degradation and in others of an improvement in the environmental situation, no evidence of an overall advance of the desert has been found in the Sahel.

Nomadic Pastoralism as a Source of Resilience

The apparent contradiction between these observations can be explained if one considers the geographical and social diversity of the region, the complex dynamics of semi-arid ecosystems, and the coevolution over several millennia of these ecosystems with the human societies that use them. The Sahel today displays great variation with regard to natural and demographic conditions, the degree to which local economies participate in regional and world commerce, the influence of national development policies, and local strategies of agricultural production.

A new ecological model that describes semi-arid ecosystems as unstable, far-from-equilibrium systems helps to explain the resilience of the Sahel in the face of drought. Even in the absence of any human impact, the plant cover changes constantly in response to random dry episodes. By way of analogy, one may think of a ball rolling around an inclined plane: the ball is always attracted by the lowest point, which corresponds to a stable position; if both the position and the tilt of the plane constantly changes before the ball has time to reach the next lowest point, however, it remains permanently far from equilibrium. Similarly, the vegetation of arid and semi-arid environments is continually disturbed by erratic variations in rainfall.

Domesticated herbivores and controlled brushfires have been part of the ecology of the Sahel for almost four thousand years. The region's ecosystems are therefore well adapted to human influence, within the limits of the modes of land use with which they have coevolved. The pastoralists of the Sahel have traditionally shown great mobility and flexibility in relation to grazing lands, following the rains to feed their herds. The ecosystems of this region have therefore developed ways to cope with this type of opportunistic land use that depend particularly on the diversity of plant species they support.

There is an optimum density of herbivore populations, for example, that maintains—and even increases—the productivity of semi-arid pastureland. A moderate level of grazing reduces the vegetative surface area from which water is lost by transpiration, and therefore indirectly increases the moisture of the soil and, with it, the potential for plant growth. This in turn stimulates plant production and favors the selection of those species most useful for livestock, with the result that the grasses become more productive, hardier, and acquire greater nutritional value than grasses in areas that have never been grazed.

In a region subject to frequent droughts where pastoral management is based on the mobility of herds, no herbivore population ever becomes large enough to degrade the pastureland, for its growth is checked by periodic fluctuations in rainfall. The traditional pastoral activities of the Sahel are therefore not to blame. What has long been thought to be desertification is due in fact principally to climatic variations and droughts, which cause temporary and reversible changes in the state of the Sahel's ecosystems, as satellite observations of the region confirm.

Semi-arid ecosystems can be degraded, however, by an unusual combination of simultaneous and mutually reinforcing climatic and human disturbances. To revert to our earlier analogy, certain abrupt movements of the inclined plane may push the ball outside the plane, causing it to fall to the ground. This pattern of events has been responsible for desertification in various parts of the world at particular junctures in their history. In the Iberian peninsula during the sixteenth and seventeenth centuries, for example, a period of political, cultural, and economic change coincided with the Little Ice Age, a climatic event that brought harsher winters to the north of Europe and wetter winters to the south. The introduction in the southeast of Spain of cereals and sheep by the Christian populations of the country's interior worked to strip the soil of its cover, exposing it to rapid erosion by the rains that accompanied the new climatic conditions. The conversion of the forests of the Pyrenees into grazing lands and the extraction of wood for naval construction led to a similar erosion of mountain soils in the northern part of the peninsula.

At the beginning of the twentieth century in North America, the cultivation of cereals was concentrated on the best soils of the Great Plains. The First World War exhausted grain stocks in Europe, causing prices to rise. In response to this economic opportunity, the area of land devoted to cereals doubled in the American Midwest between 1910 and 1920, at the expense of grazing lands. At the beginning of the 1930s, a long drought struck the region, denuding soil that had been made more fragile by deep plowing and violent winds. Storms raised up enormous clouds of dust, creating the famous "dust bowl" of the Depression. In 1933, the sky was darkened for 139 days, blocking out sunlight and therefore lowering temperatures. In May 1934, dust clouds reached New York and Washington, some fifteen hundred miles away. One day in April 1935, known thereafter as "Black Sunday," residents of small towns in the Midwest who had gone outdoors to enjoy the spring weather saw a gigantic black mass coming toward them at a speed of sixty miles per hour, preceded by a cloud of fleeing birds. Millions of acres of crops were buried under three to four inches of soil torn away by the wind from adjacent farmland.

In certain parts of Sahelian Africa, the exceptional rains of the 1950s and 1960s led to a northward expansion of small peasant farming, further encouraged by national agriculture policies, just as cattle herders were being pushed back farther north toward the desert, where pasturelands were temporarily green again. The drought that followed in the 1970s trapped the cattle herders in these desert zones, their traditional grazing routes farther south now being permanently occupied by farmers, and caused a humanitarian catastrophe in the Sahel as well as conflict between cattle herders and farmers.

These examples illustrate three possible mechanisms of desertification: climatic and socioeconomic changes may occur independently of each other at the same time (as in the Spanish case); socioeconomic change may render a society vulnerable to climatic variability (as in the American case); or a favorable climatic episode may lead a society to engage in an unsustainable activity (as in the case of the Sahel). The majority of cases of desertification in the world today, and throughout human history, have resulted from some such combination of mutually reinforcing human and climatic factors. Only an extraordinary conjunction of events has the power to push semi-arid ecosystems beyond their threshold of resilience. Satellite observations reveal that over recent decades such situations remain localized in the Sahel, even though the region as a whole suffered a sizable decline in rainfall in the 1970s and 1980s.

When Outsiders Meddle

The impression that policy makers form of the scope and causes of desertification influences the policies that they adopt. Despite the accumulation of scientific data regarding the need to maintain the mobility of herds and to manage semi-arid pasturelands in a flexible manner, as a function of fluctuations in rainfall, national leaders and international agencies have for more than three decades now implemented an inappropriate set of management policies, which, instead of combating a genuine case of desertification, has led instead to real degradation of the environment.

These interventions are founded on models of pastoral management that are poorly suited to the complex dynamics of semi-arid ecosystems, based as they are on the experience of managing large ranches in North America, where climatic conditions are more stable than in the Sahel. It is as if a patient suffering from a benign tropical condition were examined by a physician from northerly climes who is unacquainted with local epidemiology. Mistakenly believing that he recognizes the symptoms of a familiar disease, the physician tries to cure it by inappropriate methods that risk creating a more serious health problem than the initial one.

A few decades ago, when the myth of the advancing desert was widely accepted, foreign experts who blamed traditional mobility-based pastoral systems for degrading semi-arid pasturelands urged that these lands be managed in accordance with modern methods that would limit the movement of the semi-nomadic cattle herders, in particular by settling them around permanent watering places and excluding them from certain areas.

The concentration and continual presence of livestock around these watering places led to overgrazing. Farther away, farmers were encouraged to cultivate the most productive pasturelands and traditional grazing corridors, restricting still further the pastoralists' mobility. With the promotion of irrigated agriculture along rivers, cattle herds lost access to water. In certain places an attempt was even made to arrest the progress of the desert by planting a "green wall" of trees—a little like police setting up a roadblock to catch fleeing criminals—but whose effect once again was to diminish the mobility of cattle herders.

The results were beneficial to no one. Livestock were confined to areas where the risk of drought was higher. The old pasturelands now closed off to herders were impoverished owing to a profound modification of the plant cover, and tensions between farmers and cattle herders were aggravated to the point that violent confrontations occurred in certain regions. Furthermore, the intervention of the state in traditional pastoral systems weakened local societies and economies, causing a flight from the land. And because institutions that had formerly regulated access to pasturelands and watering places were no longer operative, these areas were now open to all comers, not least outside investors enjoying privileged relations with government officials.

This political response was also motivated by the fact that nomadic pastoralism is incompatible with the modernization schemes promoted by governments in the region and the related objectives of improving education and access to health services; still less does it suit state prerogatives such as border control and tax collection. The Tuareg (who move between Mali, Niger, and Algeria) and the Tubu (who move between Chad and Libya) have therefore been systematically marginalized as a matter of official policy.

An analysis limited to the physical changes wrought by environmental degradation gives only a partial view of the phenomenon, however, which affects not only vegetation and the soil, but the system formed by human society and its environment as a whole. Environmental degradation and the ability of local societies to rehabilitate degraded natural resources, while finding substitutes for them as necessary, are two sides of the same coin. Thus, for example, a drought will have a different impact on neighboring villages depending on their reserve stocks of grain, the existence of family and mutual aid networks, employment opportunities outside the agricultural sector, local programs for food assistance and irrigation development, and so on. Traditional pastoral societies have always invested a significant share of their resources in social capital, forging alliances with other groups for the purpose of guaranteeing access to distant pasturelands, rendering mutual assistance in the event of drought, ensuring security along migration corridors, and establishing both formal and informal rules for the use of pasturelands and watering places.

When the traditional mechanisms of socioeconomic resilience in response to climatic fluctuations are replaced by the purely economic expedient of importing resources from neighboring regions, new environmental and socioeconomic risks arise. Thus, for example, if the mobility of cattle herders is replaced by the importation of forage from neighboring regions in order to feed a settled population of livestock in the dry season, the customary mechanism for regulating the size of herds—low productivity of pasturelands during dry periods—is eliminated, with the result that herds grow rapidly. This phenomenon has caused serious environmental degradation both in Central Asia, under the Soviet regime, and in China. And if the provision of forage is interrupted, the pastoral system collapses—at considerable social and economic cost. This recently occurred in Kazakhstan, where the dismemberment of the Soviet Union led to the disappearance of public subsidies for animal feed.

In all of this may be seen a parable for recent human modifications of the environment on a global scale: when artificial processes and resources (the industrial system of energy production from hydrocarbons) permit a society to partially free itself from the constraints associated with natural resources and processes (the provision of wood for fuel, for example), the population has a tendency to grow rapidly and to overexploit other natural resources, on which it nonetheless remains wholly dependent (such as water and biodiversity). Economic growth accelerates. Because the provision of indispensable goods and services is unable to keep pace with the expansion of needs, however, mankind becomes more and more dependent on the artificial processes and resources it has introduced, which in turn increases its vulnerability to unanticipated disturbances that affect its ability to provide these artificial resources (political unrest in regions that hold most of the world's petroleum stocks, for example) or to disturbances that result from their use (such as climate change, caused by the burning of fossil fuels). Artificial systems seldom possess the same degree of diversity and resilience in the face of exceptional circumstances as natural systems.

A pessimist sees the difficulty in every opportunity; an optimist sees the opportunity in every difficulty.
WINSTON CHURCHILL

7 Solutions

The natural environment is changing rapidly on a global scale under the influence of human activity. These changes threaten to undermine the hope of improving the quality of life for all of the earth's people. Indeed, achieving this goal will depend on maintaining the long-term habitability of the planet.

The risk that an environmental crisis will adversely affect human societies throughout the world is now greater than ever before. In Chinese the ideogram for "crisis" consists of two characters, the first signifying danger and the second opportunity. The coming decades will be rich in technological, institutional, and cultural opportunities for putting mankind back on a sustainable path of development.

No resource has created more decisive opportunities throughout history than the human brain. The inexhaustible inventiveness of mankind, applied to the management and development of natural resources, is its most valuable asset in tackling the many environmental challenges that face the world today. Human creativity gives every reason for optimism, so long as it is used to serve the right purposes and is coupled with a heightened sense of urgency.

One way to go about deciding what needs to be done to arrest the degradation of the earth's environment is simply to enumerate its causes, on the assumption that eliminating each of them will suffice. In some cases, where a reorientation or modification of the factors responsible for an instance of environmental degradation may reasonably be expected to avert a crisis, this approach makes sense. But it will not work in all cases. It needs to be recognized that certain factors are intrinsic features of human development, and that the environmental changes to which they give rise are part of a complex process of coevolution between humanity and nature. In many cases the solution consists, not in eliminating such factors, but in reinforcing the mechanisms of mutual adaptation between human activities and ecosystems in order to achieve a more harmonious relationship among the elements that make up the terrestrial system as a whole, including the human beings that occupy it. The effectiveness of the solutions proposed therefore largely depends on the accuracy and precision with which the causes of environmental change are analyzed. Different solutions may also need to be implemented on several time scales: whereas certain solutions can have immediate effects, others demand patient work that will bear fruit only after several decades.

Opposing Schools of Thought

The general models we looked at earlier, in chapter 3, suggest some simple strategies for controlling the causes of environmental change. Certain schools of thought advocate, sometimes very aggressively, a single solution—often ignoring the fact that each environmental problem is different, and that each particular geographical and historical situation calls for a specific combination of approaches. Let us briefly review these solutions in turn.

- What might be called the "fewer mouths to feed" school advocates controlling demographic growth, especially in poor countries, where it is highest, in order to diminish the pressure exerted by population on natural resources. The reasoning is simple: if there are fewer inhabitants, a smaller number of persons will consume environmental resources, will want to drive a car, and so on. Invoking the limited carrying capacity of the planet, this school contemplates a variety of measures: the promotion of modern contraceptive methods; economic development (prosperity being itself an effective form of contraception); an increase in the survival rate of infants and children, so that poor families do not compensate for infant mortality through a high birth rate; improvement of the status of women, so that they can acquire control over their fertility; and education of both men and women. On this view, there is no "best" formula, for everything depends on the cultural and social context peculiar to each country. The debate over demographic growth is strongly influenced by religious beliefs and by the suspicion that rich countries wish to control the population of poor countries in order to preserve their privileges and their level of consumption. Yet the rate of growth of the world population is falling more rapidly than experts predicted ten years ago, and it is probable that before the end of this century (perhaps as early as 2070) world population will have stabilized and actually have begun to decline. In the course of the past fifty years, developing countries have seen a drop in fertility rates, the average number of children coming into the world having fallen from six to three per woman. In 2050 the world population should reach roughly 9 billion, and it is unlikely that it will much exceed 10 billion before beginning to taper off. Soon the world will no longer have to take care of some 77 million more children every year. But environmental change will not therefore cease to occur—all the more as the aging of the world's population will pose new challenges, particularly in Europe.

- The "bigger pic" school calls for accelerated development of new technologies and faster economic growth in order to meet the needs of a growing population by producing more. Its adherents place great faith in the power of the environmental Kuznets curve.
- The fierce (though increasingly less influential) opponents of this last tendency, who make up the "return to the candle" school, call instead for technological regression out of an unshakable belief that all technological progress necessarily increases the destruction of the environment by human activity. Below I shall discuss the possibility of striking a sensible balance between these two extreme positions by stimulating the development of new technologies whose environmental impact is low.
- The "smaller portions" school argues that mankind must consume less. A reorientation of consumption toward qualitative rather than quantitative objectives seems a more realistic prospect, however, than an actual reduction in consumption. Later in this chapter I will consider the sort of changes in cultural values that would be necessary to profoundly modify the composition of mass consumption.
- The "equal shares" school lays emphasis on the need to make the distribution of wealth and access to resources more equitable, in keeping with the moral imperative of achieving greater fairness in human opportunities. Members

of this school note that the risk of exceeding the carrying capacity of the planet comes from the fact that 20 percent of the world's population consumes a disproportionate share of natural resources. It is hardly surprising that the legitimate aspiration of the other 80 percent to escape from poverty should profoundly disturb planetary ecological equilibrium. At the same time, the great migrations of our time from the south to the north work to bring about a more balanced distribution of wealth: since the industrialized countries monopolize the fruits of growth, the inhabitants of poor countries not unnaturally come looking for their share of the pie. This school also advocates better management of the planet's resources and denounces misguided short-term policies, perverse subsidies, and corrupt practices it sees as responsible for environmental degradation and the unequal distribution of resources.

- The "tend the garden" school, which insists on the necessity of maintaining and regenerating natural capital, has given rise to a new branch of ecology devoted to developing techniques for restoring degraded ecosystems. It adopts a pragmatic approach, seeking to re-spin the web that ignorance and the lure of profit have torn, by replanting forests and reintroducing species where they have disappeared.
- The "let's privatize the pie" school regards environmental degradation as the result of a tragedy of the commons and demands that ecosystems be placed under the control of private entities. If the various elements of nature were privatized, its members argue, the natural resources to which access today is free would be protected by their owners, for human beings only take proper care of those resources to which they possess a title of ownership that excludes anyone else from using them. On this view, the appropriation of natural resources by private entities is an indispensable first step toward integrating the value of the goods and services furnished by ecosystems with the operation of markets: commercial transactions cannot apply to public goods to which access is open and therefore free of charge.

Each of these solutions focuses on one set of causes of environmental change. A radically different approach accepts that environmental change is at least to some degree inevitable, because the socioeconomic development of human societies necessarily involves the modification of the environment, and argues that, by its very nature, such change is part and parcel of the behavior of complex far-from-equilibrium systems. The resilience of human societies and ecosystems alike can be increased by appropriate measures, both by weakening positive feedbacks, in order to slow the rate of change, and by reinforcing negative feedbacks, so that ecosystems can reorganize themselves.

The advantage of this approach is that it acts upon the internal dynamics of the system rather than on the external causes of change. By promoting more adaptive forms of socioeconomic development, rather than rejecting environmental evolution, it becomes possible to channel human energies toward a sustainable path of growth.

It is important to stress that neither one of these two general approaches—acting on causes or on feedbacks—is capable by itself of achieving the right balance between natural processes and human development. As with every complex problem, only a skillfully measured and administered combination of elements of all of these solutions, with due regard for local circumstances and the peculiarities of individual cases, will bring about a successful transition to sustainable development. In what follows I briefly examine a few of the most promising ideas for technological, institutional, and cultural innovation. Once again, it needs to be emphasized that no one of them alone will be enough to meet the environmental challenges of the coming decades. All of them have already been implemented in advanced industrial societies, though for the most part only in embryonic form.

Ecoefficiency

Technological innovations make it possible to reduce the impact of human activity on the environment and, in certain cases, to restore the natural functions of the environment. There is an urgent need for technologies that increase the efficiency with which energy, land, water, and materials are used to produce goods and services, so that the same level of production can be sustained using fewer natural resources and creating less waste. The great technological innovations of past centuries increased the productivity of labor. Innovations in the decades to come must increase the productivity of resources.

Ecoefficiency can be achieved by applying to natural resources the principles that Henry Ford applied in the 1920s to labor and machines. The productivity of labor has increased by a factor of two hundred in industry since the eighteenth century: with the advent of more efficient machines and a superior organization of labor, one person is now able to do what two hundred people were needed to do three centuries ago. But productivity in the use of natural resources and energy per unit of production has enjoyed a comparatively modest increase over the same period: producing a ton of steel, for example, now requires only ten times less energy than a hundred years ago.

The potential for increasing the ecoefficiency of existing products is considerable. Automobile makers are now developing prototypes that are lighter than current models, thanks to the substitution of composite materials for steel, and more fuel-efficient, being powered by a hybrid electric/gas engine (the electricity is generated during deceleration) that consumes 70 to 80 percent less fuel than conventional gas engines of twenty years ago. Some hybrids have already been brought to market.

Reductions in the cost of energy and primary materials that flow from increased ecoefficiency boost profits, thus creating win-win situations from both the environmental and the economic point of view. At the household level, an energy-saving refrigerator reduces electric bills while diminishing the atmospheric emissions associated with the production of electricity. Downloading software from the Internet is not only faster and less expensive than buying it in a store; by eliminating the need for a physical disk, it becomes possible to dispense with the need for packaging, and so to avoid the pollution caused in shipping it to retail outlets by truck. The recovery of methane gas from garbage dumps provides a low-cost fuel from an abundant, renewable, and cheap resource: waste. Once the cost of investment in clean technologies has been absorbed, reductions in natural resource consumption and pollution hold out the promise of increased prosperity.

The ecoefficiency of most production operations can be substantially improved, especially in less developed regions of the world. There is still a large gap between the global average rate of productivity for most industrial activities and the productivity made possible by more effective—and currently available—technologies and methods. For example, the average agricultural yield for corn is four tons per hectare worldwide, as against an average of seven tons in the United States (rising to seventeen tons in Iowa and twenty-one tons in the case of irrigated fields). To be sure, such yields will never be achieved in ecologically marginal regions, and overly intensive, poorly managed agriculture is liable to generate serious environmental problems anywhere. But these figures do suggest that the general adoption of already familiar technologies would permit a considerable increase in the productivity of natural resources. What is more, the most advanced technologies presently available are not capable of reaching theoretically maximum levels of productivity, calculated on the basis of insurmountable physical limits.

An increase in resource productivity is only one of a number of possible ways to relieve environmental pressures. Investment in natural capital and a systematic attempt to limit loss and waste in the processing and distribution of food products, for example, also help save resources. Feeding the world's population may require not so much increasing the quantity of agricultural inputs (putting more land under cultivation, for instance, or using more fertilizer and water) as making more efficient use of resources that are already available (by optimizing the use of fertilizers, farmland, and irrigation, and by preserving the quality of soils and water). Moreover, on average, 15 percent of agricultural production is lost after harvesting in the course of storage, treatment, and transport. This figure exceeds 40 percent in certain countries in Africa, where food shortages are most acute.

Three imperatives need to be taken into account, then, if natural resources are to be properly managed: the diffusion of existing technologies, development of new technologies, and reduction of waste. But the positive effects of ecoefficiency in its economic dimension may lead to a rebound effect. Because reducing production costs makes products more affordable, it also stimulates consumption, so that the savings from lower costs are apt to be swallowed up by the increase in consumption. The same thing happens when the widening of a highway, rather than serving to improve traffic flow, ends up attracting more motorists. A similar effect can be observed on the social level as well, as a new appreciation of the beauty of nature leads a growing number of families to build homes in wooded rural areas, rather than to protect these areas. Consuming more products—even if now they do somewhat less harm to the environment than they did in the past—is hardly a way to advance the cause of sustainable development. This rebound effect poses a problem for the boundless faith placed by optimists in the capacity of technology to solve environmental problems.

Decarbonization

Another challenge to technological innovation is to find new ways to further "decarbonize" energy systems. The burning of biomass or hydrocarbons not only releases energy; it also produces pollutants, creating smog (a combination of fog and air pollution) and climate change through a strengthening of the greenhouse effect (as a consequence of the emission of carbon dioxide). And yet the most useful element in hydrocarbons for the production of energy is not carbon, but hydrogen.

For two hundred years now, industrialized societies have progressively diminished the carbon content of their energy supplies. From the chemical point of view, the ratio between the number of carbon and hydrogen atoms in the most common fuels has steadily declined. Wood, the principal fuel before the Industrial Revolution, contains about ten useful carbon atoms for every hydrogen atom: its combustion releases high quantities of carbon dioxide. Coal approaches parity (one or two atoms of carbon per atom of hydrogen). Petroleum contains on average two hydrogen atoms for every carbon atom. A molecule of methane (CH_4) is made up of four atoms of hydrogen per carbon atom. Nuclear, hydroelectric, solar, and wind energies produce no carbon dioxide at all.

Between 1860 and 2000, the volume of carbon used in the production of primary energy fell on average by 0.3 percent a year per unit of energy produced—a decline of more than 40 percent over the entire period. The promise of further decarbonization offered by hydrogen in the form of fuel cells, and the development of renewable nonpolluting power sources such as eolian energy, suggests that this trend will continue.

Eolian energy already satisfies 20 percent of the electric needs in Denmark and 5 percent in Germany (as against only 0.12 percent in the United States in 1999). These figures could rise to 30 percent by linking wind farms together in a network, in order to compensate for unproductive days at individual sites. (The new turbine engines used in wind farms today, with their longer blades and lower rotation speeds, no longer have fatal consequences for bird populations as early models once did.) The electricity generated in this way could be used to make hydrogen (from water), which could then be stored in fuel cells to power cars and other machines. Further development of renewable energies will require that they be subsidized on the same scale as energies derived from oil and coal, however, with national programs put in place for constructing the necessary infrastructure.

More efficient use of energy in daily life also contributes to the decarbonization of industrialized economies. On a global scale, the quantity of energy used per unit of gross national product (GNP) reached a peak around 1925, falling to almost half that level by 1990. Today additional savings can be achieved in a number of ways, for example, by switching to more economical electric household appliances and compact fluorescent light bulbs (which consume five times less energy than normal bulbs and last ten times as long), reducing energy loss during its transmission and distribution through the power grid, and by adopting energy-saving architectural innovations (such as new materials with high thermal inertia, which retain the solar energy received during the day and release it at night; improved insulation; south-facing windows; and so on). Reductions in energy consumption per unit of GNP have already been achieved in a great many countries owing to the economies of scale permitted by consolidating energy production and to the growth of the service sector at the expense of industry, a greedier consumer of energy.

The American entrepreneurs and ecologists Paul Hawken, Amory Lovins, and L. Hunter Lovins argue that success in improving energy efficiency depends on applying the "lobster principle": one eats the large pieces of meat from the tail and the claws first, and then, with dexterity and patience, searches for meat in the crevasses of the shell and in the legs. In other words, deep and lasting energy economies can only be achieved by paying attention to the details: once the major sources of waste have been eliminated, it is necessary to identify and exploit the many small possibilities for reducing consumption that remain.

These developments are far from sufficient, however, for the total consumption of energy and the total emissions of carbon on a planetary scale have increased together with the growth of the world economy. Energy needs per unit of economic production and carbon emissions per unit of energy consumed have declined, but world economic production has grown by a greater amount; since the Industrial Revolution, energy consumption has increased on average by more than 2 percent per year (world energy consumption in 1990 was seventy-five times higher than in 1800).

Dematerialization

Technological innovations that diminish the amount of materials needed for industrial purposes can also have positive consequences for the environment. "Dematerializing" the economy reduces the need for primary materials, lessens the impact of industrial production on the landscape, lowers the risks to health associated with the use of toxic materials (cadmium, lead, and so on), and reduces the quantity of waste products. With regard to primary materials, each unit of GNP in industrialized countries since the beginning of the twentieth century has required less and less wood, steel, copper, and lead, but more and more plastic, aluminum, phosphate, and potassium (the last two being important components of chemical fertilizers). The substitution of plastic for metal products led to a reduction in the weight of the materials used, but also to an increase in their volume.

A number of industrial products—such as personal computers, mobile telephones, and portable digital media players—are now smaller and lighter than before and contain more recycled materials. As in the case of decarbonization, however, the tendency to dematerialization is still in its early stages, and there are further opportunities for making products lighter as well as for miniaturization and recycling. All the elements necessary for profoundly reforming industrial production are at hand: improved product efficiency and longevity, better design using smaller quantities of higher-quality materials, salvage of the waste generated in the course of production, reuse and recycling of used products. Consumers stand to gain from this development as much as the environment.

But here again the increase in consumption of finished products offsets and, in fact, exceeds the gains achieved by dematerializing production. In other words, despite the smaller quantity of materials used per unit of production, the overall consumption of materials increases (in the United States even the total consumption of wood doubled during the course of the twentieth century). Instead of reducing per capita consumption of materials, the miniaturization of household appliances and their increasingly multipurpose use have led to the acquisition of still more appliances. Each family member wants to have his own car, his own mobile telephone, his own music player, his own television set, his own personal computer—whereas only a short time ago, one of each of these items per family seemed ample. The quantity of materials per unit of volume has sharply declined in the construction of houses, but houses are now larger than they used to be, even though households are growing smaller. Cars are built with less steel than in the past, but consumers now buy bigger cars (sport utility vehicles to pick up the children from school). Airplanes are lighter, but they fly more frequently and farther than they used to. The Internet gives access to vast quantities of electronic information, but its users print out search results. Mailboxes are overflowing with advertising flyers that no one reads. Per capita consumption of paper has never been higher than it is today.

The Battle against Pollution

Production technologies have become less polluting as well. The air in the cities of industrialized countries is safer to breathe than it used to be. In the United States, where between 1940 and 1990 emissions of atmospheric pollutants fell by an average of 3 percent per unit of production, the reduction in pollution has been more rapid than the increase in overall production. In the major cities of the United States and Europe, atmospheric pollution from sulfur dioxide and carbon monoxide is today only about a third of what it was thirty years ago. Concentrations of fine particles, nitrogen oxides, volatile organic compounds, and ozone have diminished from 50 percent to 25 percent (though these pollutants still pose risks for human health and ecosystems alike, and their degree of concentration in some large cities continues to regularly exceed the limits considered acceptable for health). The concentration of lead in the air in wealthy countries has fallen by 90 percent where unleaded gasoline is used, and urban smog is much less common than it was. Some rivers are cleaner than they used to be in wealthy countries as well. A salmon was caught in the Rhine in 1992, the first since its congeners were driven out of the river by pollution more than two centuries earlier. Salmon reappeared in the Thames in 1974 after an absence of 140 years.

This progress has been due to hundreds of technological innovations, promoted and supported by government policies, that have achieved greater ecoefficiency in transportation (catalytic converters in exhaust systems, improved engines, better gas), energy production (coal gasification, mixed-cycle gas turbines, wind farms, and possibly underground storage of carbon dioxide, now in the experimental stage), construction (reduced energy loss, use of recycled materials, increased energy efficiency of electric household appliances and heating systems, lower fluorocarbon emissions), industry (reduced toxic waste, use of new materials, improvements in energy management, increased energy efficiency of industrial operations), and waste management (recycling, use of methane from monitored garbage dumps to produce heat and electricity).

These technical advances have been made more rapidly, for the most part, than experts had predicted in the early 1990s. But they probably would not have occurred at all in the absence of targeted environmental policies that created new opportunities for innovation. Further research and testing is needed to speed up the development of such technologies and promote their widespread adoption, while ensuring that their environmental impact continues to be low.

For the moment, however, these advances remain the privilege of wealthy countries. Poor countries seldom have access to new technologies and lack the economic capacity to devote research resources to their development. In the major cities of these countries—home to some 2 billion people, or about a third of the world's population—atmospheric pollution has already reached disturbing levels and grows worse by the year. The World Health Organization estimates that air pollution contributes to the death of more than 5 million children in the world annually.

Technology sometimes has the effect of substituting one environmental problem for another. Before the car replaced the horse in large cities, tons of manure had to be disposed of. The major source of urban pollution in those days was this foul-smelling dung, a carrier of disease that attracted swarms of flies. Additionally, the carcasses of dead horses had to be removed by the thousands. Almost five acres of grain were needed to feed each horse, which contributed to an expansion of farmland at the expense of forests on the outskirts of cities. In 1900, England counted one horse for every ten inhabitants and the United States one for every four inhabitants.

With the introduction of the car, roads became cleaner but the quality of the air worsened. On a global scale, atmospheric emissions associated with transportation increased at an accelerated rate. Pollution since the era of horse-drawn transport has therefore been redirected away from city streets to the atmosphere, but also toward coastlines (in the event of oil spills) and oil-rich regions.

Plans for Restoration and Repair

New technologies are now being developed to repair the damage caused to the terrestrial system by human activity. A new field called restoration ecology has made it possible to hasten the recovery of ecosystems, for example, by reproducing natural processes as far as possible by artificial means, on the basis of detailed studies of the functioning of the ecosystem where intervention is contemplated.

At the other extreme, an approach known as geoengineering seeks to use advanced technologies to manipulate the terrestrial environment on a vast scale. Pioneered by the Soviet Union in the 1930s, and later used for military purposes—without great success—by the United States during the Vietnam War, geoengineering has been proposed as a solution to the problem of global climate change due to human activity since the 1980s. The idea is not to eliminate the causes of climate change or to adapt to its effects, but instead to alter the climate system as a whole. Proposals include sending gigantic mirrors into space that would reflect the rays of the sun away from the earth, thus counteracting global warming; fertilizing the oceans with iron in order to stimulate the growth of plankton, which absorbs atmospheric carbon; creating genetically modified organisms capable of absorbing carbon both on land and in the sea; and, still more speculatively, modifying the ocean currents (which distribute heat around the globe) by the construction of giant barriers. These proposals, all of them the object of serious research, reflect a boundless confidence in the technological prowess of modern societies. The possible perverse—and potentially irreversible—effects of such interventions are enough to make one shudder.

The Role of Institutions

As we have seen, technological innovation is a necessary, but not by itself a sufficient condition for solving environmental problems. It must be accompanied by organizational and cultural changes as well. To attain a path of sustainable development, institutions—the set of rules, decision-making procedures, and programs that shape social practices—must be transformed in order to stimulate technological innovation, implement already known technological solutions on a large scale, and counteract the rebound effect described earlier.

Here the challenge is to create an institutional environment that encourages private actors to assume short-term costs in order to create long-term benefits for the community, through a skillful mix of positive incentives to protect the environment with the threat of negative sanctions in the case of degradation of a public environmental good; of state intervention with appeals for voluntary action on the part of private businesses; and of legislative tools with market mechanisms. The proper balance can be achieved only within an efficient system of governance that reserves a large measure of freedom to private actors.

In order to diminish the adverse impact of economic activities on the environment, institutions can make use of the legal system to encourage private actors by means of a combination of incentives and sanctions to respect the environment. These may include laws defining environmental norms, punishing destructive practices, regulating firms' right of entry in certain sensitive sectors (by granting or withholding licenses to exploit natural resources), prohibiting products and materials that pose an environmental danger, and setting ecoefficiency standards.

This top-down approach needs to be supplemented by a bottom-up approach, so that citizens' groups are able to sue private industries or the state, claiming damages and interest for harm suffered in the event that their health or property is injured by pollution in one form or another. Following the example of former smokers who successfully brought lawsuits against the tobacco industry in the United States, persons who have suffered damages from environmental causes are now able to look to the courts for various kinds of remedy and relief.

In 2002, for example, nine states in the northeast United States jointly sued the Bush administration for having weakened provisions of the Clean Air Act, a major piece of environmental legislation passed by Congress under the Nixon administration. The complaint charged that the Bush administration's decision to no longer require offending factories to take steps to control pollution was responsible for acid rain, smog, asthma, and respiratory diseases that would affect millions of Americans. Another suit was recently brought by state attorneys general against the Bush administration for not having taken action to arrest global climate change. No matter that the value of such actions is largely symbolic, the American example shows that a country can take steps to modify its own environmental laws, compensate the victims of environmental injury, and ensure legal representation for whistle-blowers who take polluters to court.

Legislation placing undue burdens upon businesses can have unintended effects, however. Thus, for example, excessive production costs arising from the need to comply with strict environmental rules risk reducing the competitiveness of individual firms and therefore their capacity to invest in new technologies. In the worst case, such laws can lead companies to relocate their operations in countries with more relaxed environmental standards—a way of exporting pollution from rich countries to poor countries.

More than two thousand years ago, in the *Tao Te Ching*, Lao Tsu observed that "governing a large country is like frying a small fish: one must be careful not to poke it too much." No one disputes that environmental policies have an impact on the competitiveness of firms, but there is disagreement as to whether the impact is negative or positive. While strict environmental standards may add to the costs of production, they also encourage firms to modernize their operations and improve their ecoefficiency, and in this way gain a competitive edge in markets that will inevitably become increasingly regulated with regard to various forms of pollution. A firm's reputation for respecting the environment can become a reason for loyalty on behalf of its customers, employees, and shareholders as well. Consumers are increasingly inclined to vote with their pocketbooks, boycotting businesses whose practices they consider environmentally unsound and favoring products whose ecological fitness has been certified, even if they cost slightly more. Investors flock toward socially responsible mutual funds that bring together companies with good records on environmental and social issues, in addition to turning in strong financial performances.

Institutions and the Market

Institutions can also intervene in the markets in order to encourage economic actors to behave in ways that minimize their impact on the environment. States can intervene by means of subsidies designed to promote environmentally sound practices, taxes on pollution and waste ("ecotaxes"), certification procedures for assessing the quality of environmental management, and the creation of new markets for services furnished by natural ecosystems (markets for water purification, for example, or for protecting biodiversity in the form of land under natural vegetation cover).

In each case, it is necessary to find the environmentally most efficient formula while leaving the greatest possible freedom to producers and consumers. For example, a tax on the carbon content of fuels is more efficient than a tax on their energy content: an energy tax penalizes all forms of energy consumption, whereas a carbon tax favors clean fuels.

The possibilities for promoting sustainable production and consumption are limitless, as Hawken and the Lovinses have shown in *Natural Capitalism* (1999). For example, if architects' fees were based in part on energy savings achieved through the use of new heating and lighting techniques, they would have a strong incentive to search for the most ecoefficient solutions. To pay heating engineers on the basis not of the number of boilers installed, but of the number of Fahrenheit-degree hours supplied to a building, would encourage them to look for ways to furnish this quantity of heat at the least cost. They would prefer high-efficiency boilers, collaborate with thermal insulation companies, and develop a long-term relationship with clients for the maintenance and improvement of heating systems.

For certain goods, the practice of leasing encourages the repair and recycling of obsolete machinery by producers, who retain ownership throughout the lifetime of their equipment and therefore have an interest in designing it in such way that its parts can easily be reused at the end of its life. Imposing a road tax on cars and calculating automobile insurance premiums as a function of the number of miles traveled would encourage drivers to restrict the use of their vehicle and implement the "polluter pays" principle.

More fundamentally, if taxes on labor and capital were partially converted into taxes on resources and pollution, businesses would have an interest in reducing their consumption of nonrenewable resources and lowering pollution emissions by using a larger workforce. New workers would be given the task of recycling waste and parts. Engineers would specialize in the search for energy savings and the development of new production techniques intended to promote dematerialization. Additional research would be directed to making resources more productive and production processes less polluting. In this case, the resolution of environmental problems would go hand in hand with an increase in employment.

A simple political decision to modify the tax structure—while maintaining a constant volume of taxation—would suffice to send firms the signal that their host country is resolutely committed to achieving a sustainable path of development. Pollution and unemployment are not inevitable outcomes of industrial development: they result from inappropriate taxation policies and a set of distorted incentives that work against labor and in favor of environmental degradation.

Of course, transforming the tax system requires that great care be taken so that the fiscal burden is equitably distributed among the various sectors of the economy and the various classes of society. A tax system can only be changed gradually, since firms must have a reasonable amount of time to adapt. The sole purpose of fiscal reform should be to align the prices of products with their real costs, including the cost that their production and consumption places upon the environment, and therefore upon future generations.

The Creation of Environmental Markets

The assignment of tradable emission quotas (or pollution permits, as they are often called) is another way of allowing firms to satisfy their environmental obligations at low cost, perhaps even profitably.

Such a market for pollution has operated in the United States since the early 1990s, when the Environmental Protection Agency began distributing emission permits for sulfur dioxide, which is responsible for urban air pollution and acid rain, to electricity producers throughout the country. This amounted to giving them a title of ownership for certain levels of sulfur dioxide emissions. Producers are prohibited from polluting beyond a specified threshold, under threat of sanctions. Since the objective is to reduce pollution, this quota is set at a level lower than the quantity of pollution emitted during a base year. Producers are able to buy and sell their emission permits.

The cleanest and most ecoefficient companies manage to reduce their emissions to a level below the quotas that have been assigned to them, at a relatively low cost, and then sell their surplus pollution permits to companies that pollute the most. By contrast, the companies whose technology and equipment are the least ecoefficient would have to invest considerable sums to reduce their emissions: buying pollution permits costs them less than to modernize their plants. In more technical terms, companies with high marginal pollution reduction costs buy pollution permits from companies whose marginal reduction costs are low. The price of these permits rise or fall in accordance with supply and demand. A futures market has emerged as well, in which speculators buy pollution permits in the expectation that their value will increase over time.

A few years after this program was put into effect, the desired reduction of sulfur dioxide emissions had been achieved, at a cost ten times less than had initially been estimated and without excessively rigid and constraining rules. The challenge now is to lower pollution quotas, a move that will not go unopposed. Some environmental economists propose extending this arrangement to cover services furnished by ecosystems, such as carbon fixation in forests, water purification, flood prevention, plant pollination, the renewal of soil fertility, and the preservation of biodiversity.

The fairness of a "cap and trade" system of this sort depends largely on the way in which permits are granted and natural resources are exploited. Environmental markets involve the distribution of what originally were public—and therefore free—goods to private entities. This privatization of nature, as it were, raises troublesome questions of fairness. Who has the right to distribute permits to catch a certain quantity of migratory fish in the earth's oceans? Who has the right to receive these permits? Who can participate in the exchange of these permits? Would it be acceptable if a fishing fleet from a rich country were to buy up all the fishing permits granted to the companies of a poor country, even if these permits covered the right to fish in the territorial waters of the poor country? How would one prevent a fleet from buying the permits of its competitors, without any intention of using them, for the sole purpose of bringing about a rise in prices?

Clearly there is no choice but to regulate environmental markets in such a way that privatized natural resources do not become concentrated in the hands of a few, who would thereby acquire a monopoly on services furnished by the earth's ecosystems. Existing antitrust legislation guarding against the abuse of conventional forms of market power would have to be extended to environmental markets.

The concept of tradable pollution permits is applied today within certain large corporations, where an internal market has been created to allow different divisions to achieve energy savings targets. It has also been implemented on a global scale within the framework of the Kyoto Protocol on climate change. In this case, the permits concern the emission of carbon, which, in the form of carbon dioxide, contributes to the greenhouse effect. A Brazilian farmer who reforests his plot of land, for example, thus enabling a small quantity of carbon in the atmosphere to be absorbed by the vegetation, can sell his carbon credits on an international market to an industrialized country that has not managed to fulfill its treaty obligations in reducing carbon dioxide emissions. In this way the polluting country will be able to decrease emissions slightly less than it would otherwise be obliged to do, the farmer will increase his income, the pace of reforestation will be accelerated, and the level of carbon dioxide in the atmosphere will be reduced.

Again, there are problems that will need to be worked out. One crucial question has to do with the procedures for monitoring and verifying attempts to protect the services of natural ecosystems, particularly over the long term. How are we to ensure, for example, that newly planted stands of trees are not burned to make room for farmland or destroyed by an uncontrollable forest fire, once the owners have sold their carbon credits?

Even so, the most remarkable aspect of these initiatives, as the American ecologists Gretchen Daily and Katherine Ellison have pointed out, is that the search for financial gain, which is partly responsible for the degradation of the environment, can be productively channeled toward the resolution of environmental problems. By modifying the rules of the game, the pursuit of individual advantage can be reconciled with the long-term public good: creating markets for pollution permits means that the services provided by ecosystems are no longer free—their value is captured by these environmental markets and reflected in their price. In this way, both negative externalities (pollution and degradation of ecosystems that affect a great number of actors) and positive externalities (the benefits that accrue to everyone through the protection of nature) are internalized.

It is necessary, then, to establish a system of prices and payments that accurately reflect the impact—negative and positive alike—of human activity on ecosystems. Only once the true value of the services they furnish is incorporated in daily decisions will nature be protected. In the formula of the American economist Richard Sandor, instead of considering nature as a fixed-price "all you can eat" buffet—for in this case few of us can pretend that we would not overindulge—we should regard it instead as an à la carte menu.

Some militant environmentalists are bound to find the notion of environmental markets disturbing. But by making it profitable to protect nature, it becomes possible to attract entrepreneurs and capital to nature-conservation activities—something that militancy has so far been unable to achieve, except in the case of a few charitable organizations.

International Agreements

Coordinated actions on an international scale, in the form of multilateral treaties, seek to ensure that the burden of protecting the natural environment is equitably shared. In the course of recent decades, more than a hundred conventions on the environment have been signed with regard to matters as diverse as the ozone layer, fishing, climate change, desertification, biodiversity, wetlands, migratory species, and so on. Most of these treaties suffer from the same defects: interminable negotiations, vague commitments, divergent interpretations, lack of coordination between agreements, inadequate provisions, nonratification by the worst offenders, bureaucratic complexity in application, and insufficient enforcement mechanisms. Most of the time, it is necessary to count on national or regional implementation of rules negotiated on an international scale. Some critics charge that such agreements are little more than dead letters, having only symbolic value.

One of the essential purposes of these treaties is to encourage the diffusion of advanced, low-pollution technologies to poorer countries through investment, transfers of technical expertise, financing arrangements, and the training of local experts. After all, if for every new car made with the currently most advanced technology that is purchased in the European Union a used car is sent to Albania or Senegal, no global environmental problem is solved. Optimists may claim that economic development has the effect of reducing pollution, but this is true only of wealthy countries. Only major investments on the part of these countries will enable developing countries—and therefore the world—to follow a path of sustainable development.

All the solutions proposed here depend on the will of governments to create a favorable institutional context and to overcome the resistance of pressure groups that defend particular interests. The example of the Scandinavian countries shows how rapidly governmental intervention can transform the way that the environment is managed, without hurting economic performance. The counterexample of the United States under George W. Bush is an unsettling reminder that the difference of a few hundred votes can result in the dismantling of several decades of effective environmental legislation, with grave consequences for the entire planet and for future generations.

Cultural Responses

If everyone in the world were to become vegetarian by the end of this century, it would be possible to feed a global population of 9 billion people using current agricultural technology but without increasing the area of land presently under cultivation. Fully one-third of world grain production today goes to feed livestock: several pounds of grain are required to produce a pound of meat. These figures go to show that the environmental impact of human activity depends to a large extent on the mode of life of the planet's inhabitants—what they consume, how they live—and not only on their technologies of production and their institutions. The consumption of goods and services is the end point of a whole chain of production decisions. Ultimately, it is the choices of consumers that determine what is produced, in what quantities, and by what means.

The increase in domestic comfort, leisure time, and the availability of material goods has progressively enlarged the impact of human societies on the environment over recent centuries. Compared to their ancestors, people today possess more spacious and better-heated homes, accumulate more objects to furnish and decorate them with, travel greater distances, and devote a larger part of their time and money to recreational activities that consume energy and natural resources. These changes in people's ways of living have more than offset the gains in ecoefficiency due to new technologies.

Ways of living change as a function of social and demographic characteristics such as income, time spent working and pursuing leisure activities, household size, life expectancy, and length of retirement. Apparently insignificant factors such as office and store schedules influence the energy consumption of firms and the commuting habits of individuals: extended business hours increase the length of time that buildings need to be heated and lighted, and, by spreading out road and highway use over time, encourage more people to rely on their own car for transportation. More fundamentally, ways of living depend on the evolution of social norms (customs) and individual preferences (tastes), which jointly determine these sociodemographic factors.

The real challenge facing wealthy countries is to find a way to break the link between quality of life and the unbridled consumption of material goods and energy-intensive services (in developing countries, an increase in consumption is necessary in order to alleviate the poverty that affects a large part of the population). As we have seen, the urge to consume can never be entirely satisfied. The need for particular goods can be fulfilled, however, and consumption reoriented toward other, less environmentally harmful kinds of goods.

The only example of a satisfied need in the world today is the desire for food among members of high-income social groups, since there is a limit to the amount of food that can be consumed in a day by a healthy person. Just the same, the generic need to eat can evolve from a quantitative interest in eating more to a qualitative interest in eating better (a healthier and more varied diet that relies on organic ingredients and makes it possible to prepare more flavorful dishes, and so on). Improving the quality of meals need not entail an increase in the environmental impact of consumption; indeed, when organic products are used, the impact may be lower.

Looking to the future, other generic needs come to mind that could be satisfied by engaging in environmentally preferable activities: covering short distances on bicycle for exercise and, wherever possible, walking rather than driving; establishing a sense of personal identity and social status by donating time and money to worthy causes and taking advantage of cultural and educational opportunities, rather than by seeking to possess costly material goods; finding self-fulfillment and the satisfactions of social life by taking part in local efforts to preserve and restore nature (the creation of a protected area in one's neighborhood, for example) rather than by making long trips by car to watch events that degrade the environment (such as sports car racing); investing in energy-efficient improvements to one's home rather than enlarging it; and so on.

The American geographer Robert Kates argues that changes in consumption must be promoted on the basis of six principles, which affect both supply and demand: reducing the energy and materials content of goods (decarbonizing energy and dematerializing manufactured goods); shifting consumption toward goods that fulfill the same needs as before but whose negative environmental impact is lower (promoting renewable energies); substituting information-rich goods for ones that make large demands on energy and materials (pursuing a vibrant social and cultural life); deepening the enjoyment one finds in what one already possesses, through more intensive use of previously acquired goods (taking more time to enjoy one's garden, for example); resisting the temptation to re-create needs that have already been satisfactorily met (holding on to one's present car rather than trading it in for a new model); and sublimating the need for material goods with higher-level needs (seeking to refine one's social, moral, and aesthetic sensibilities).

A new style of consumption that considers more to be too much carries with it a moral imperative, namely, that there be enough for everyone.

Equity in the international distribution of wealth is intimately linked to changes in the patterns of consumption in wealthy countries. The search for a new model of consumption based on the quality of life, rather than the quantity of material goods, is not incompatible with economic growth; to the contrary, it amounts to advocating the growth of a service sector whose environmental impact is low.

This change in consumption habits, if it is to influence the behavior of the majority of people, can only come about over the long term, for it must go against a powerful tide of incessant appeals to constantly consume more. Of all the scientific, technological, and social challenges posed by the impact of human activity on the environment, modifying the urge to consume is probably the most complex. On the one hand, formidable commercial pressures condition the most vulnerable segments of the world's population—children, adolescents, and certain classes of people in developing countries. Deconditioning them is not easily done (as we know in the case of tobacco, despite the fact that the tobacco industry cannot claim that cigarettes satisfy a fundamental, healthy human need). On the other hand, the deep motivations underlying consumption, and the ways in which it is shaped by the characteristic individualism of Western culture, are still poorly understood.

Modifying the social perception of certain types of consumption is nonetheless not a utopian scheme. Consider the transformation in Western societies of the image of the smoker, from the strong and independent figure of the 1960s to the compulsive personality of today, sadly unable to control his self-destructive urges. A similar change in the image of the overconsumer and polluter can be imagined during the course of the next few decades. Indeed, it has already begun to occur on an international scale with the Bush administration's decision in 2001 not to sign the Kyoto Protocol, which created a widespread perception of a selfish, go-it-alone government committed, by its devotion to a few blinkered industrial lobbies, to ensuring short-term prosperity for a minority of the world's population without regard for the cost to future generations everywhere on earth.

Changing habits of mass consumption requires a transformation of individual attitudes toward the purposes of social life: each person must learn to see the world as something more than a gigantic supermarket and to reconsider the place of mankind on the earth. As the American economist Richard Norgaard observes, until now the most important human decisions have been made on the assumption that what is good for mankind is good for the planet. From now on, the growing interactions between human activity and the terrestrial system require a complementary principle that enjoins us to act as if what is good for the planet will be good for us as well.

In the twentieth century, the system of social values and organization was modeled on technological development, in particular on the exploitation of fossil energy sources and the industrial processes that the abundance of energy made possible. But technology, though it has its own logic, must now be seen only as a means, not as an end in itself; as being beneficial only if it serves a higher social ideal. Moreover, changes in collective rules and individual behaviors will have a much more decisive impact on the environment over the long term than changes arising from technological innovation alone. It is our duty to forge a new system of values based on a shared sense of belonging to one and the same world, and a related sense of social responsibility.

A Few Simple Ideas

Imagine a law that requires drivers, whenever there is room in their vehicle, to pick up hitchhikers. You could leave your house and get in the first car that comes along, even if it means having to change vehicles once or twice to arrive at your destination, without ever having to wait. An equilibrium would automatically be established between the number of cars on the road and the number of hitchhikers. When this latter number is low in relation to the number of cars, waiting time would be minimal, which would make hitchhiking an attractive option. An excess of hitchhikers would encourage a greater number of people to drive. Under this regime, a person who prefers to use his own car assumes the full cost of the trip: by giving a ride to a few people, he pays the community, in the form of a service, for the cost that society incurs through his use of his car.

This measure would end up costing society less than what it pays now in the form of pollution, which would also be lowered, for hitchhiking would reduce traffic by a factor of two or three (assuming that each car carries two or three passengers). It would have many other benefits as well: diminished gas consumption, reduced pollution, fewer traffic jams, and more efficient use of roads and automobiles—to say nothing of its contribution to a deeper sense of community. Would such a policy infringe upon individual liberty? No matter that each car is a piece of private property—the roads belong to the state, which therefore has the right to lay down the rules of their use.

So far as I know, this idea has never been applied in its simplest form. Carpooling, an organized and institutionalized version of it, involves only a fraction of the population. In California, for example, cars crossing the Bay Bridge between Berkeley and San Francisco are exempted from the toll charge if they carry several passengers. Alternative ways of getting around can be developed as well. The city of Pasadena distributes bicycles to municipal employees who agree to use them instead of cars. In Copenhagen, public bicycles can be picked up at various locations throughout the city for a nominal security deposit (the equivalent of about four dollars), and reused by other riders once they have been returned to any bicycle parking lot (the principle is identical to the one governing the use of shopping carts in supermarkets, only applied on the scale of the city as a whole). Riding a bicycle in Copenhagen has become a symbol of social status and an expression of solidarity, flexibility, and responsibility toward the environment.

Albert Einstein remarked that a way of thinking that has created a problem is powerless to solve it. This explains why complex and costly technological solutions to environmental problems, which often are only an extension of the very technologies that are responsible for such problems, are so seldom effective. One alternative approach, which grows out of the work of the Oxford-trained German economist E. F. Schumacher, searches for solutions that lie outside the logic of a particular problem. "Thinking outside the box" is not easy, but it often leads to simple solutions, obvious once they are identified, and relatively inexpensive.

Some governments in Sahelian Africa used to encourage the purchase of gas stoves in the hope of replacing wood as the principal domestic fuel and limiting deforestation. Since few peasants could afford to buy propane gas tanks, this costly technological solution achieved little. In the 1980s a nongovernmental organization went around villages promoting the manufacture and use of small, very simple ovens made from baked clay known as *foyers améliorés* (improved fireplaces) that considerably increased the fraction of heat produced by burning wood that is actually used (the amount of wood consumed is 30 percent or less than the amount required by traditional methods of cooking with a casserole set on top of stones). Villagers learned to construct this stove out of locally available materials, at negligible cost. This rudimentary solution has proved to be more effective in reducing rates of deforestation than any other.

Despite the fact that high-quality water resources are becoming increasingly scarce, the same water that is used for drinking and cooking in industrialized countries continues to be used for flushing toilets, watering gardens, and cleaning cars. The existence in every home of two water systems, one containing potable water, the other rainwater collected on rooftops, would permit considerable savings. More simply still, repairing leaking toilets, faucets, and pipes would allow each household to save on average a tenth of its water consumption.

The best inventions are the simplest ones. Indeed, the American physicist and inventor Edwin Land used to say that an invention is nothing more than "a sudden cessation of stupidity." Agriculture, to cite only one example, can reduce its water needs by as much as 95 percent by not inundating fields for the entire growing season. Rather than watering crops at full pressure in the middle of the day, when the sunlight causes a high rate of evaporation, it is less wasteful to bring water directly to the roots, drop by drop, when the soil's moisture falls below a certain threshold (detected by an underground sensor). Yet only a few years ago, precision irrigation systems of this type were in use on only 1 percent of the world's irrigated land.

Moreover, the best inventions solve several problems at once: instead of concentrating on particular problems that affect isolated compartments of a system of production and consumption, they address the needs of the system as a whole. The city of Curitiba in Brazil has set up a program of "green exchange" in disadvantaged neighborhoods, where vouchers for public transport or food (eggs, milk, oranges, and potatoes purchased by the city from nearby farms) are issued in exchange for bags of garbage sorted by the residents. Thus, for example, two pounds of recyclable waste may be traded for one pound of food. In this way it has been possible to recycle two-thirds of Curitiba's waste. In addition to keeping neighborhoods clean, assuring a healthy diet for the city's poorest residents, supplying small businesses with recyclable material, creating employment, and avoiding the spread of diseases associated with water pollution, this program guarantees a stable demand for agricultural products, which keeps farmers productively occupied in their fields rather than languishing in shantytowns.

Response Time

The *Titanic* sank not because its captain did not see the iceberg that the ship was headed toward, but because he saw it too late. Whereas technological progress had considerably increased both the ship's mass and its cruising speed, and therefore the time needed to change course, the technology for detecting drifting icebergs had not kept pace. Similarly, one may wonder whether our ability to anticipate environmental risks has evolved as rapidly as our ability to create them. The challenge facing humanity today is incomparably greater than the challenges facing small societies in the past that were forced during environmental crises to modify their path of development. Considering their size, growing complexity, and absence of central direction, it is plain that modern social and economic systems have acquired an extraordinary degree of inertia.

It may not be enough, however, to point out dangers looming on the horizon. History shows that human societies have a disturbing tendency to wait until threats become tangible before reacting, which increases the risk of catastrophe. From the late nineteenth century onward, the city of London suffered from frequent smog, caused by the smoke from factories, domestic coal-burning stoves, and, in the twentieth century, diesel-burning buses. In 1873, the pollution was so dense that pedestrians, unable to find their way, actually fell into the Thames. The impact of smog on health had been widely studied during the Victorian era, but it was not until a sad week in December 1952, during which a thick, black, sulfurous fog caused the death of four thousand people (and almost double this number in the three months that followed), that scientific knowledge was translated into effective political action. Even so, a clean air act was not approved by the British Parliament until 1956, the Conservative government of the day having in the meantime tried to blame the deaths on an influenza epidemic (Great Britain was then a very profitable exporter of its own coal, keeping the dirtiest grade, laden with sulfur, for use in electric power stations and homes).

Similarly, reinforcement of the greenhouse effect by atmospheric pollution was first reported by the Swedish chemist Svante Arrhenius in 1896. One hundred years later, no concrete political measure had yet been enacted. Worse still, the present American government persisted until very recently in casting doubt upon the reality of climate change—while generously subsidizing the extraction of coal and oil. Must history be allowed to repeat itself, so that a major climatic accident has to occur before effective countermeasures are instituted? The difference in this case is that the whole of the planet is at risk, not just a single city.

The problem of making decisions under conditions of uncertainty is as old as the world itself. But if necessary decisions are put off for too long, with irreversible consequences, the cost of delay will turn out to be exorbitant. Today, paradoxically, the very governments that have embraced a strategy of preventive action in foreign policy and military affairs refuse to adopt it in the only domain where it is easily justified, namely, the protection of the environment.

Even if action is prompt and effective, some current conditions will persist in modified form. The ozone layer will take a century to be reconstituted, assuming that destructive chemical substances are no longer expelled into the atmosphere. The carbon dioxide that was added to the atmosphere in the twentieth century will remain there for more than a century, during which time it will continue to warm the climate. The pollution of closed seas such as the Mediterranean will linger for decades, even if pollutants are no longer dumped into it on a daily basis. It will take several centuries for eroded soils in certain parts of the world to be restored. In some regions, overexploited populations of fish will require several decades to regenerate once fishing has ceased. Extinct plant and animal species will never reappear. The Aral Sea will remain a salt desert for decades.

It needs to be kept in mind, too, that it takes several decades for a society to absorb the implications of a major technological change, above all if the various technologies involved are mutually dependent and have different replacement rates. Automobiles, although they replaced horses after only a few years, took half a century to become widely used in Western countries: the development of a modern automobile culture depended on long-term investments in infrastructure (paved roads and highway systems), a modification of the human habitat (suburbanization), and an increase in incomes.

Similarly, the principal obstacle to the widespread adoption of the hydrogen-powered car is not so much the time needed to replace the stock of conventional automobiles as the cost of creating the infrastructure for storing and distributing hydrogen. Automakers will not market hydrogen cars until 20 to 30 percent of existing service stations are equipped to sell hydrogen; and service stations will not make the necessary investments until a sufficient number of hydrogen cars are on the road.

The diffusion of ideas and institutions is still slower. A society needs about a century to fundamentally change its cultural values. Given the enormous inertia of modern technological, institutional, and cultural systems, it is essential that humanity react rapidly to the environmental challenges that rise up before it today. In the words of the American biologist Peter Vitousek: "We are the first generation with the tools to understand changes in the earth's system caused by human activity, and the last with the opportunity to influence the course of many of the changes now rapidly underway."

Never despair. Infuse more.
HENRI MICHAUX

Conclusion

Mankind has always reorganized natural systems, and environmental changes do not necessarily degrade the environment. Such changes may represent a degradation for some users of an ecosystem and an improvement for others, depending on their purposes and time horizons. In trying to decide whether an environmental change is positive for a group of users, we need to take into account the following factors: the value of the goods and services provided by the modified ecosystem (it must be higher, for the population of users as a whole, than the value of goods and services supplied by the ecosystem in its natural state); the series of long-term consequences that follow from human modification of a natural resource (taking into account whether or not resources are renewable, their rate of regeneration, cascading effects throughout the ecosystem, and so on); the consequences for individuals who are not directly involved (an environmental change is positive only if the negative externalities are minimal); the fairness of all such consequences, for both present and future generations (since most of the services furnished by nature are public goods, all users have an equal right to benefit from these services); and the need to ensure, as scientific knowledge increases, that the present use of natural ecosystems will keep future options open and preserve a minimal stock of intact natural resources (an extinction of biological species, for example, eliminates the possibility that they might be employed one day for benign human purposes—for example, by furnishing genetic material to create new pharmaceutical products).

Throughout human history, many societies have managed to reverse a trajectory of environmental degradation by making cultural, political, social, economic, and technological changes. Others, however, have collapsed, particularly when the ruling class has not perceived, anticipated, or understood the nature of this degradation, or else has not reacted; or when the technological, institutional, and economic capacity for innovation has been lacking. As we saw in chapter 5, a variety of conditions concerning information, motivation, and capacity must be fulfilled for a transition to sustainability to succeed.

At the present moment in its history, as I say, humanity has embarked upon a vast experiment for which there is no central guidance, no long-range plan, no possibility of turning back, and no second chance. Its current path of economic development, if followed without any major modification, is untenable. In view of the planetary scope and the unprecedented rapidity of contemporary environmental changes, the inertia of natural and social systems, and the growing complexity of globalized economies, humanity must anticipate future crises and move quickly to adjust the terms of its relationship to the environment accordingly. The mode of development preferred by modern societies until now has relied upon creative destruction—a sequence of events whereby the destructive exploitation of natural resources leads to a crisis, which in turn stimulates the introduction of new modes of exploitation. The risks of this strategy are increased when changes occur rapidly and on a global scale. Once certain thresholds have been exceeded, alterations of the environment are liable to be irreversible, rendering new, belatedly introduced modes of resource management inoperative.

Another mode of development is possible. After all, as a Saudi oil minister once exclaimed, our ancestors did not emerge from the Stone Age because they ran out of stones! What may be called adaptive development rests on a more dynamic view of the interactions between human activities and the natural environment. It seeks to strengthen the ability of societies to deal with environmental change, and in this way to increase their resilience in the face of unwelcome surprises. Societies that adapt successfully anticipate environmental problems by regularly evaluating how well ecosystems are functioning; they detect problems at an early stage and solve them before they become serious.

This ability to continually make corrections to the course of economic and environmental development assumes an efficient political system in which all actors participate in the making and implementation of decisions; that is, a truly democratic system, in which the great majority of citizens share a common vision of the future, a vision that mobilizes individual enthusiasm and creativity. Like the tightrope walker and the monkey perched on his shoulder, the earth and the people who live on it are precariously suspended over an abyss. Adaptive development is the strategy that the monkey must now adopt: it cannot wait until the tightrope walker comes within an inch of falling to stop jumping about; instead it must coordinate its movements with those of its host, so that they advance along the tightrope together, as though they were one, ceaselessly reacting, rapidly and skillfully adjusting their progress to a seemingly endless series of small oscillations that threaten to pull them fatally far away from equilibrium.

When all is said and done, must we be pessimists or optimists with regard to the future of our planet—and therefore the future of humanity? A qualified optimism is in order: we can be optimists on the condition that the pessimism justified by current scientific knowledge forces us from now on to modify our ways of living, far more significantly and more rapidly than we have done until now. No insurmountable economic or technological barrier blocks our way. The future depends on our clear-sightedness and on the choices we make, both as individuals and as a body politic. The best way to predict the future is to invent it. To do this, societies must carefully weigh a series of fundamental options.

Markets vs. Regulation

Can market mechanisms solve environmental problems, or is state intervention necessary? Does the sum of individual decisions necessarily lead to a long-term common good?

The market is the most efficient and most flexible means of persuading economic actors to join in the search for solutions and of lowering costs. But there are strong reasons to believe that regulation of markets by the state is indispensable if the true value of the goods and services supplied by nature is to be incorporated in economic decision making. State intervention is needed also to lay down the rules by which the private good and the common good can be made to converge.

Centralization vs. Decentralization

Is a vast world government necessary to manage the environmental problems of the planet, or should decisions be delegated to lower levels of authority, where a body of knowledge adapted to local circumstances can be more readily found and where cohesive social networks exist?

As we have seen, a concerted effort by governments at several levels, which coordinate their actions and exchange the necessary information (but no more than this), is the most appropriate solution, because it permits decisions to be taken at the hierarchical level at which a given problem arises and where the competence to solve it exists. This principle of subsidiarity places the burden of responsibility upon those who have the greatest interest in finding a solution to the problem and maximizes the overall efficiency of the system. Each subsystem, though it is autonomous and responsible for its own behavior, can turn to higher levels for assistance.

Reform vs. Revolution

Can the challenges of sustainable development be met by a series of small reforms, or is revolution the only alternative?

The strategy of gradual reform tends to promote "end of pipe" solutions—small innovations that reduce the inconveniences of an existing technology while reinforcing its dominant position and prolonging its life. Rather than invest in hydrogen-powered cars and a comprehensive public transportation network, the exhaust systems of gas-powered cars are equipped with catalytic converters. Rather than invest in new industrial processes, the smokestacks of polluting factories are raised so that they disperse their toxic fumes over a larger territory. These small solutions minimize short-term costs but delay a change that is nonetheless inevitable, and one that will be all the more expensive for society as it is pushed back further into the future.

More radical changes, though they have the advantage of solving certain problems, risk imposing high costs upon business and eliminating firms that are incapable of transforming their operations and investing in cleaner technologies, with potentially destabilizing socioeconomic consequences. It is therefore necessary to enact the right policies at the right moment in each area of environmental policy, in each region of the world. There are times that call for incremental change, which keeps the system close to equilibrium when the knowledge needed to successfully carry out thoroughgoing reforms is lacking; there are other times when only profound change will do, for delay will have costly—indeed, irreversible—economic and environmental consequences. Profound change must in any case be initiated calmly and deliberately, however, on the basis of original and sound ideas that command widespread support.

Prevention vs. Reparation

When human activity has a negative impact on the environment, is it necessary to remove the cause of the degradation (for example, by prohibiting the pollutants responsible for the destruction of the ozone layer)? Or is it preferable to develop new technologies capable of compensating for the damage, even of repairing it (for example, by injecting industrially manufactured ozone into the stratosphere)?

As a general rule, reparative intervention should be undertaken only when the functioning of the system in question has been perfectly understood, so that unintended secondary effects can be avoided.

Domination vs. Adaptation

Can humanity expect one day to be able to dominate natural processes, or must it adjust to their ebb and flow? Can it overcome the constraints imposed by nature and succeed in eliminating the variability of natural processes, or must it increase its adaptive capacities?

The current state of our understanding of the terrestrial system is far too fragmentary for mankind to pretend that it is capable of mastering nature. Living and coevolving with nature is our only option. Every tsunami, every hurricane, every earthquake reminds us—even those of us who live in advanced industrial countries—that nature will not allow itself to be tamed.

The major challenge in the decades ahead is to create a world rich in win-win situations, one that is rich in opportunities for improving the quality of life in our societies and narrowing social and economic inequalities, while at the same time reducing the pressure exerted by human activity on the terrestrial system. Where such opportunities do not exist, we must define our priorities and be ready to renounce short-term enrichment in favor of long-term development.

But change depends also on our attitude as individuals. The scale and complexity of environmental problems often arouse a sense of impotence, the impression that the weight of personal decisions is negligible, and the conviction that the future course of events will be decided by forces larger than ourselves. Yet if 6.5 billion people were to modify their daily decisions—even if only slightly, but nonetheless in convergent directions—a profound revolution in the history of humanity would occur.

Every one of us is at once a consumer, a voter, an educator, and, at least potentially, an innovator. As consumers, we possess the power to vote with our pocketbook: to choose a product for its environmental performance in addition to its quality; to pay attention to the ecological certification or labeling of certain products, such as tropical wood; to boycott brands whose owners have furnished proofs of chronic negligence in relation to the environment. More important still, we can control our consumption of material goods and determine how much is enough. We can modify our preferences for goods that consume large amounts of material and energy, and devote more of our time to social and cultural activities whose effects are environmentally positive or otherwise neutral. We can simplify our lives and choose to concentrate on the things that matter most to us.

As voters, we have the freedom to vote for political candidates and parties that campaign on behalf of sustainable development. Green parties do not have a monopoly on such concerns, and the spectrum of possible solutions extends well beyond the traditional cleavages between left and right. Between elections, it is our duty to remind our representatives of their prior commitments, to confront them with their contradictions, and to communicate our priorities to them.

As educators—particularly of our children, grandchildren, and students—we play a decisive role, helping to transmit values that go beyond our immediate personal interest, among them a sense of responsibility toward the world and toward future generations, a willingness to subordinate our individual desires to the general welfare, and a commitment to act in a way that benefits humanity and the planet on which we live, and which we will bequeath to those who come after us.

As innovators, we can continually improve our management of resources, correct our errors, share lessons learned, and show ourselves receptive to new thinking, whatever its source. Simple ideas, put into effect by a great many people, are likely to have a greater impact than a few expensive gadgets reserved for an elite few.

But any technological innovations and cultural revolutions that may take place in advanced industrial economies will have a negligible impact on the future of the planet if they do not also satisfy the needs of the 80 percent of the world's population who live in developing countries: this part of humanity, marginalized during most of the twentieth century, is now making its presence felt in the great game of economic growth, with all of the environmental consequences that this implies. To ensure that it follows a path of development that avoids the waste and environmental destruction characteristic of our own industrial evolution during the past two centuries will require unprecedented levels of investment and a worldwide sharing of knowledge and technologies.

The ideas and cultural impetus needed to make sustainable development a reality are not lacking, even if for the time being they affect only a minority of the world's population. Helping to inaugurate a new way of living on the planet Earth—in agreement with its diversity, its complexity, and its beauty—is the responsibility of each one of us. What is more, it is within the power of each one of us.

Bibliography

Preamble

Carson, Rachel. *Silent Spring*. Boston: Houghton Mifflin, 1962.

Introduction

Abbott, David R., ed. *Centuries of Decline During the Hohokam Classic Period at Pueblo Grande*. Tucson: University of Arizona Press, 2003.

Alcamo, Joseph, et al. *Ecosystems and Human Well-Being: A Framework for Assessment*. Washington, D.C.: Island Press, 2003.

Balmford, A., A. Bruner, P. Cooper et al. "Economic Reasons for Conserving Wild Nature." *Science* 297 (2002): 950–53.

Caldwell, L. K. "Is Humanity Destined to Self-Destruct?" *Politics and the Life Sciences* 18 (1999): 3–14.

Condorcet, Jean-Antoine-Nicolas de Caritat. *Esquisse d'un tableau historique des progrès de l'esprit humain*. Paris, 1795.

Crutzen, P. J. "Geology of Mankind." *Nature* 415 (2002): 23.

Daily, G. C., ed. *Nature's Services: Societal Dependence on Natural Ecosystems*. Washington, D.C.: Island Press, 1997.

Jablonski, D. "Extinctions: A Paleontological Perspective." *Science* 253 (1991): 754–57.

Lomborg, Bjørn. *The Skeptical Environmentalist: Measuring the Real State of the World*. Cambridge: Cambridge University Press, 2001.

Malthus, T. R. *An Essay on the Principle of Population; or, A View of Its Past and Present Effects on Human Happiness*. 8th ed. London: Reeves and Turner, 1878.

McNeill, J. R. *Something New Under the Sun: An Environmental History of the Twentieth-Century World*. New York: W. W. Norton, 2001.

Myers, N. "Environmental Unknowns." *Science* 269 (1995): 358–60.

Norgaard, R. B. "Optimists, Pessimists, and Science." *Bioscience* 52 (2002): 287–92.

Ponting, Clive. *A Green History of the World: The Environment and the Collapse of Great Civilizations*. New York: St. Martin's Press, 1991.

Redman, Charles L. *Human Impact on Ancient Environments*. Tucson: University of Arizona Press, 1999.

Steffen, W., A. Sanderson, P. D. Tyson, J. Jäger, P. Matson et al. *Global Change and the Earth System: A Planet Under Pressure*. Berlin: Springer, 2004.

Tainter, Joseph A. *The Collapse of Complex Societies*. Cambridge: Cambridge University Press, 1988.

Chapter 1

Achard, F., H. D. Eva, H. J. Stibig, P. Mayaux, J. Gallego et al. "Determination of Deforestation Rates of the World's Humid Tropical Forests." *Science* 297 (2002): 999–1002.

Ball, J. B. "Global Forest Resources: History and Dynamics." In *The Forests Handbook*, vol. 1., edited by J. Evans, 3–22. Oxford: Blackwell Science, 2001.

Belyaev, A. V. "Water Balance and Water Resources of the Aral Sea Basin and Its Man-Induced Changes." *GeoJournal* 35 (1995): 17–21.

Chow, J., R. J. Kopp, and P. R. Portney. "Energy Resources and Global Development." *Science* 302 (2003): 1528–31.

Crutzen, P. J. "Geology of Mankind." *Nature* 415 (2002): 23.

Dirzo, R., and P. H. Raven. "Global State of Biodiversity and Loss." *Annual Review in Environment and Resources* 28 (2003): 137–68.

Dixon, J. A., L. M. Talbot, and G. J. M. LeMoigne. *Dams and the Environment: Considerations in World Bank Projects*. World Bank Technical Paper 110. Washington, D.C.: International Bank for Reconstruction and Development, 1989.

Döös, B. R. "Population Growth and Loss of Arable Land." *Global Environmental Change: Human and Policy Dimensions* 12 (2002): 303–11.

Fagan, Brian M. *How Climate Made History, 1300–1850*. New York: Basic Books, 2001.

Folke, C., Å. Jansson, J. Larsson, and R. Costanza. "Ecosystem Appropriation by Cities." *Ambio* 26 (1997): 167–72.

Food and Agricultural Organization (FAO). "Fisheries: Databases and Statistics." http://www.fao.org/fi/statist/statist.asp, 2003.

———. *Global Forest Resources Assessment 2000 (FRA 2000): Main Report*. FAO Forestry Papers 140. Rome: Food and Agricultural Organization, 2001.

Glazovsky, N. F. "The Aral Sea Basin." In *Regions at Risk: Comparisons of Threatened Environments*, edited by J. X. Kasperson, R. E. Kasperson, and B. L. Turner, 92–139. Tokyo: United Nations University Press, 1995.

Gleick, P. H. "Soft Water Paths." *Nature* 418 (2002): 373.

———. "Water Use." *Annual Review in Environment and Resources* 28 (2003): 275–314.

Goldewijk, K. K. "Estimating Global Land Use Change Over the Past 300 Years: The HYDE Database." *Global Biogeochemical Cycles* 15 (2001): 417–34.

Gregg, J. W., C. G. Jones, and T. E. Dawson. "Urbanization Effects on Tree Growth in the Vicinity of New York City." *Nature* 424 (2003): 183–87.

Grübler, Arnulf. "Technology." In *Changes in Land Use and Land Cover: A Global Perspective*, edited by W. B. Meyer and B. L. Turner, 287–328. Cambridge: Cambridge University Press, 1994.

———. *Technology and Global Change*. Cambridge: Cambridge University Press, 1998.

Hawken, Paul, Amory Lovins, and L. Hunter Lovins. *Natural Capitalism: Creating the Next Industrial Revolution*. Boston: Little, Brown, 1999.

Hutter, B. "Compliance and Beyond." *European Business Forum*. Special issue on Sustainable Development (2002): 11–12.

Intergovernmental Panel on Climate Change (IPCC). *Climate Change 2001: The Scientific Basis*. Contribution of Working Groups I and III to the Third Assessment Report of the Intergovernmental Panel on Climate Change. Cambridge: Cambridge University Press, 2001.

International Council for Science (ICSU). *New Genetics, Food, and Agriculture: Scientific Discoveries, Social Dilemmas*. http://icsudqbo.alias.domicile.fr, 2003.

International Tank Owners Pollution Federation, Ltd. *Tanker Oil Spill Statistics*. http://www.itopf.com/stats.html, 2000.

Irion, R. "The Melting Snows of Kilimanjaro." *Science* 291 (2001): 1690–91.

Johnson, N., C. Revenga, and J. Echeverria. "Managing Water for People and Nature." *Science* 292 (2001): 1071–72.

Kalnay, E., and M. Cai. "Impact of Urbanization and Land-Use Change on Climate." *Nature* 423 (2003): 528–31.

Lambin, E. F., H. Geist, and E. Lepers. "Dynamics of Land Use and Cover Change in Tropical Regions." *Annual Review of Environment and Resources* 28 (2003): 205–41.

Maddison, Angus. *The World Economy: A Millennial Perspective*. Paris: Organisation for Economic Co-operation and Development, 2001.

Matson, P. A., W. J. Parton, A. G. Power, and M. J. Swift. "Agricultural Intensification and Ecosystem Properties." *Science* 277 (1997): 504–9.

McNeill, J. R. *Something New Under the Sun: An Environmental History of the Twentieth-Century World*. New York: W. W. Norton, 2001.

Mittermeier, R., C. G. Mittermeier, P. R. Gil, J. Pilgrim, G. Fonseca et al., eds. *Wilderness: Earth's Last Wild Places*. Chicago: University of Chicago Press, 2003.

Myers, R. A., and B. Worm. "Rapid Worldwide Depletion of Predatory Fish Communities." *Nature* 423 (2003): 280–83.

Nepstad, D. A., A. Veríssimo, A. Alencar, C. Nobre, E. Lima et al. "Large-Scale Impoverishment of Amazonian Forests by Logging and Fire." *Nature* 398 (1999): 505–8.

Osborn, T. J., and K. R. Briffa. "The Spatial Extent of 20th-Century Warmth in the Context of the Past 1200 Years." *Science* 311 (2006): 841–44.

Pauly, D., and R. Watson. "The Last Fish." *Scientific American* (July 2003): 43–47.

Pimm, S. L., G. J. Russell, and J. L. Gittleman. "The Future of Biodiversity." *Science* 269 (1995): 347–50.

Population Division, Department of International Economic and Social Affairs, United

Nations Secretariat. *World Urbanization Prospects: The 2001 Revision* (ESA/P/WP.173). New York: United Nations Publications, 2002.

Ramankutty, N., and J. A. Foley. "Estimating Historical Changes in Global Land Cover: Croplands from 1700 to 1992." *Global Biogeochemical Cycles* 13 (1999): 997–1027.

Ramankutty, N., J. A. Foley, and N. J. Olejniczak. "People on the Land: Changes in Global Population and Croplands During the Twentieth Century." *Ambio* 31 (2002): 251–57.

Sala, O. E., F. S. Chapin, J. J. Armesto, E. Berlow, J. Bloomfield et al. "Global Biodiversity for the Year 2100." *Science* 287 (2000): 1770–74.

Schiermeier, Q. "How Many More Fish in the Sea?" *Nature* 419 (2002): 662–65.

Serageldin, I. "World Poverty and Hunger: The Challenge for Science." *Science* 296 (2002): 54–58.

Steffen, W., A. Sanderson, P. D. Tyson, J. Jäger, P. Matson et al. *Global Change and the Earth System: A Planet Under Pressure*. Berlin: Springer, 2004.

Tilman, D. "Global Environmental Impacts of Agricultural Expansion: The Need for Sustainable and Efficient Practices." *Proceedings of the National Academy of Sciences USA* 96 (1999): 5995–6000.

Turner, B. L., R. E. Kasperson, W. B. Meyer, K. M. Dow, D. Golding et al. "Two Types of Environmental Change." *Global Environmental Change* 1 (1990): 14–22.

Vitousek, P. M., H. A. Mooney, J. Lubchenco, and J. M. Melillo. "Human Domination of Earth's Ecosystems." *Science* 277 (1997): 494–99.

Vörösmarty, C. J., and D. Sahagian. "Anthropogenic Disturbance of the Terrestrial Water Cycle." *Bioscience* 50 (2000): 753–65.

Warren-Rhodes, K., and A. Koenig. "Escalating Trends in the Urban Metabolism of Hong Kong: 1971–1997." *Ambio* 30 (2001): 429–38.

The World Conservation Union (IUCN). *Species Extinction*. http://www.iucn.org/news/mbspeciesext.pdf, 2003.

Young, Anthony. "Is There Really Spare Land? A Critique of Estimates of Available Cultivable Land in Developing Countries." *Environment, Development, and Sustainability* 1 (1999): 3–18.

———. *Land Resources: Now and for the Future*. Cambridge: Cambridge University Press, 1998.

Chapter 2

Bertallanfy, Ludwig von. *General System Theory: Foundations, Development, Applications*. Rev. ed. New York: Braziller, 1968.

Blaikie, Piers M., and Harold Brookfield. *Land Degradation and Society*. London: Methuen, 1987.

Bossel, Hartmut. *Earth at a Crossroads: Paths to a Sustainable Future*. Cambridge: Cambridge University Press, 1998.

Costanza, R. "Visions, Values, Valuation, and the Need for an Ecological Economics." *Bioscience* 51 (2001): 459–68.

Daily, Gretchen C., and Katherine Ellison. *The New Economy of Nature: The Quest to Make Conservation Profitable*. Washington, D.C.: Island Press, 2002.

Diamond, Jared M. *Guns, Germs, and Steel: The Fates of Human Societies.* New York: W. W. Norton, 1999.

Edgerton, Robert B. *Sick Societies: Challenging the Myth of Primitive Harmony.* New York: The Free Press, 1992.

Fullerton, D., and R. Stavins. "How Economists See the Environment." *Nature* 395 (1998): 433–34.

Greenwood, B., and T. Mutabingwa. "Malaria in 2002." *Nature* 415 (2002): 670–72.

Holling, C. S., ed. *Adaptive Environmental Management and Assessment.* New York: Wiley-Interscience, 1978.

Kauffman, Stuart A. *The Origins of Order: Self-Organization and Selection in Evolution.* New York: Oxford University Press, 1993.

Marsh, George Perkins. *The Earth as Modified by Human Action.* New York: Scribner, Armstrong, 1874.

Marten, G. G. *Human Ecology: Basic Concepts for Sustainable Development.* London: Earthscan, 2001.

Masood, E., and L. Garwin. "Costing the Earth: When Ecology Meets Economics." *Nature* 395 (1998): 426–27.

Norgaard, Richard B. *Development Betrayed: The End of Progress and a Coevolutionary Revisioning of the Future.* London: Routledge, 1994.

Patten, Chris, Tom Lovejoy, John Browne, Gro Brundtland, Vandana Shiva, and HRH The Prince of Wales. *Respect for the Earth: Sustainable Development.* London: Profile Books, 2000.

Pearce, D. "An Intellectual History of Environmental Economics." *Annual Review in Energy and the Environment* 27 (2002): 57–81.

Pearce, David W., and R. Kerry Turner. *Economics of Natural Resources and the Environment.* Baltimore: Johns Hopkins University Press, 1989.

Rambo, A. T. *Conceptual Approaches to Human Ecology.* Research Report 14. Honolulu: East-West Environment and Policy Institute, 1983.

Rappaport, R. A. "Maladaptations in Social Systems." In *The Evolution of Social Systems,* edited by J. Friedman and M. J. Rowlands, 49–87. Pittsburgh: University of Pittsburgh Press, 1978.

Redman, Charles L. *Human Impact on Ancient Environments.* Tucson: University of Arizona Press, 1999.

Steward, Julian Haynes. *Theory of Culture Change: The Methodology of Multilinear Evolution.* Urbana: University of Illinois Press, 1955.

Toynbee, Arnold Joseph. *A Study of History.* 2 vols. New York: Oxford University Press, 1947–57.

Chapter 3

Arrow, K., B. Bolin, R. Costanza, P. Dasgupta, C. Folke et al. "Economic Growth, Carrying Capacity, and the Environment." *Science* 268 (1995): 520–21.

Barrett, G. W., and E. P. Odum. "The Twenty-First Century: The World at Carrying Capacity." *Bioscience* 50 (2000): 363–68.

Berkes, F., and C. Folke, eds. *Linking Social and Ecological Systems: Management Practices and Social Mechanisms for Building Resilience*. Cambridge: Cambridge University Press, 2002.

Boserup, Ester. *The Conditions of Agricultural Growth: The Economics of Agrarian Change Under Population Pressure*. London: Allen and Unwin, 1965.

Cohen, Joel E. *How Many People Can the Earth Support?* New York: W. W. Norton, 1995.

———. "Population Growth and Earth's Human Carrying Capacity." *Science* 269 (1995): 341–46.

Commoner, B. "A *Bulletin* Dialogue on 'The Closing Circle': Response." *Bulletin of Atmospheric Science* 28 (1972): 42–56.

Costanza, R. "Visions, Values, Valuation, and the Need for an Ecological Economics." *Bioscience* 51 (2001); 459–68.

Dietz, T., E. Ostrom, and P. C. Stern. "The Struggle to Govern the Commons." *Science* 302 (2003): 1907–12.

Dixon, J. A., L. M. Talbot, and G. J. M. LeMoigne. *Dams and the Environment: Considerations in World Bank Projects*. World Bank Technical Paper 110. Washington, D.C.: International Bank for Reconstruction and Development, 1989.

Ehrlich, P. R., and J. P. Holdren. "The Impact of Population Growth." *Science* 171 (1971): 1212–17.

Ekins, P. "The Kuznets Curve for the Environment and Economic Growth: Examining the Evidence." *Environment and Planning* 29 (1997): 805–30.

Ezzati, M., B. H. Singer, and D. M. Kammen. "Towards an Integrated Framework for Development and Environment Policy: The Dynamics of Environmental Kuznets Curves." *World Development* 29 (2001): 1421–34.

Feeny, D., F. Berkes, B. J. McCay, and J. M. Acheson. "The Tragedy of the Commons: Twenty-two Years Later?" *Human Ecology* 18 (1990): 1–19.

Hardin, G. "The Tragedy of the Commons." *Science* 162 (1968): 1243–48.

Holling, C. S. "Resilience and Stability of Ecological Systems." *Annual Review in Ecology and Systematics* 4 (1973): 1–23.

Kessler, J. J. "Usefulness of the Human Carrying Capacity Concept in Assessing Ecological Sustainability of Land Use in Semi-Arid Regions." *Agriculture, Ecosystems, and Environment* 48 (1994): 273–84.

Kuznets, S. "Economic Growth and Income Inequality." *American Economic Review* 45 (1955): 1–28.

Laris, P. "Burning the Seasonal Mosaic: Preventative Burning Strategies in the Wooded Savanna of Southern Mali." *Human Ecology* 30 (2002): 155–86.

Lee, R. D. "Malthus and Boserup: A Dynamic Synthesis." In *The State of Population Theory*, edited by D. Coleman and R. Schofield, 96–130. Oxford: Basil Blackwell, 1986.

Levin, S. A., S. Barrett, S. Aniyar, W. Baumol, C. Bliss et al. "Resilience in Natural and Socioeconomic Systems." *Environment and Development Economics* 3 (1998): 223–44.

Mather, A. S., C. L. Needle, and J. Fairbairn. "Environmental Kuznets Curves and Forest Trends." *Geography* 84 (1999): 55–65.

Ostrom, E., J. Burger, C. B. Field, R. B. Norgaard, and D. Policansky. "Revisiting the Commons: Local Lessons, Global Challenges." *Science* 284 (1999): 278–82.

Pearce, David W., and R. Kerry Turner. *Economics of Natural Resources and the Environment*. Baltimore: Johns Hopkins University Press, 1989.

Ponting, Clive. *A Green History of the World: The Environment and the Collapse of Great Civilizations*. New York: St. Martin's Press, 1991.

Redman, Charles L. *Human Impact on Ancient Environments*. Tucson: University of Arizona Press, 1999.

Sneath, D. "Ecology: State Policy and Pasture Degradation in Inner Asia." *Science* 281 (1998): 1147–48.

Tainter, Joseph A. *The Collapse of Complex Societies*. Cambridge: Cambridge University Press, 1988.

Turner, B. L., R. E. Kasperson et al. "A Framework for Vulnerability Analysis in Sustainability Science." *Proceedings of the National Academy of Sciences USA* 100 (2003): 8074–79.

Vellinga, P. "Industrial Transformation: Towards Sustainability in Production and Consumption Processes." *IHDP Update* 4 (2001): 6–8.

Vitousek, P. M., P. R. Ehrlich, A. H. Ehrlich, and P. A. Matson. "Human Appropriation of the Products of Photosynthesis." *Bioscience* 36 (1986): 368–73.

Wackernagel, M., N. B. Schulz, D. Deumling, A. Callejas Linares, M. Jenkins et al. "Tracking the Ecological Overshoot of the Human Economy." *Proceedings of the National Academy of Sciences USA* 99 (2002): 9266–71.

Waggoner, P. E., and J. H. Ausbel. "A Framework for Sustainability Science: A Renovated IPAT Identity." *Proceedings of the National Academy of Sciences USA* 99 (2002): 7860–65.

Wilkinson, Richard G. *Poverty and Progress: An Ecological Model of Economic Development*. London: Methuen, 1973.

Chapter 4

Abbott, Lawrence. *Quality and Competition: An Essay in Economic Theory*. New York: Columbia University Press, 1955.

Alcamo, Joseph, et al. *Ecosystems and Human Well-Being: A Framework for Assessment*. Washington, D.C.: Island Press, 2003.

Barbier, E. B. "The Economic Determinants of Land Degradation in Developing Countries." *Philosophical Transactions of the Royal Society of London* 352 (1997): 891–99.

Bourdieu, Pierre. *La Distinction: Critique sociale du jugement*. Paris: Éditions de Minuit, 1979.

Bruggink, J. J. C. "The Global Potential for Drastic Reduction of Greenhouse Gas Emissions." *Change* 50 (2000): 10–15.

Carvalho, G. O., D. Nepstad, D. McGrath, M. del C. Vera Diaz, M. Santilli et al. "Frontier Expansion in the Amazon: Balancing Development and Sustainability." *Environment* 44 (2002): 35–45.

Costanza, R. "Government-Sponsored Perversity." *Bioscience* 51 (2001): 408–10.

Cotta, Alain. *La Société ludique: La vie envahie par le jeu*. Paris: Grasset, 1980.

Crumley, C. L., ed. *Historical Ecology: Cultural Knowledge and Changing Landscapes*. Santa Fe, N.M.: School of American Research Press, 1994.

Dietz, T., and E. A. Rosa. "Rethinking the Environmental Impacts of Population, Affluence, and Technology." *Human Ecological Review* 1 (1994): 277–300.

Ehrlich, Paul R. *The Population Bomb*. New York: Ballantine, 1968.

Herodotus. *The Histories*. Book I. Trans. Aubrey de Sélincourt, rev. with introductory matter and notes by John Marincola. New York: Penguin, 1996.

International Road Federation. *White Paper on Roads and the Greenhouse Effect*. http://www.irfnet.org/archives/documents/WhitePapers/Greenhouse_interactive.pdf, 2001.

Kates, R. W. "Population and Consumption: What We Know, What We Need to Know." *Environment* 42 (2000): 10–19.

Keilman, N. "The Threat of Small Households." *Nature* 421 (2003): 489–90.

Kempton, W., and C. Payne. "Cultural and Social Evolutionary Determinants of Consumption." In *Environmentally Significant Consumption: Research Directions*, edited by P. C. Stern et al., 116–23. Washington, D.C.: National Academy Press, 1997.

Keynes, John Maynard. "Economic Possibilities for Our Grandchildren." In *Essays in Persuasion*; reprinted as volume 9 of *The Collected Writings of John Maynard Keynes*. London: Macmillan, 1972.

Lambin, E. F., and H. Geist. "Regional Differences in Tropical Deforestation." *Environment* 45 (2003): 22–36.

Lambin, E. F., H. Geist, and E. Lepers. "Dynamics of Land Use and Cover Change in Tropical Regions." *Annual Review of Environment and Resources* 28 (2003): 205–41.

Lele, U., V. Viana, A. Veríssimo, S. Vosti, K. Perkins, and S. A. Husain. *Brazil, Forests in the Balance: Challenges of Conservation with Development*. Washington, D.C.: International Bank for Reconstruction and Development, 2000.

Lindqvist, C., ed. *Globalization and Its Impact on Chinese and Swedish Society*. Stockholm: Forskningsrâdsnämnden, 2000.

Liu, J., G. C. Daily, P. R. Ehrlich, and G. W. Luck. "Effects of Household Dynamics on Resource Consumption and Biodiversity." *Nature* 421 (2003): 530–33.

Malthus, T. R. *An Essay on the Principle of Population; or, A View of Its Past and Present Effects on Human Happiness*. 8th ed. London: Reeves and Turner, 1878.

Maslow, H. "A Theory of Human Motivation." *Psychological Review* 50 (1943): 370–96.

McNeill, J. R. *Something New Under the Sun: An Environmental History of the Twentieth-Century World*. New York: W. W. Norton, 2001.

Meadows, D. H., D. L. Meadows, J. Randers, and W. W. Behrens III. *The Limits to Growth: A Report for the Club of Rome's Project on the Predicament of Mankind*. New York: Universe Books, 1972.

Mol, A. P. J. *Globalization and Environmental Reform: The Ecological Modernization of the Global Economy*. Cambridge, Mass.: MIT Press, 2001.

Myers, N. "Consumption: Challenge to Sustainable Development." *Science* 276 (1997): 53–55.

Myers, Norman, and Jennifer Kent. *Perverse Subsidies: How Tax Dollars Can Undercut the Environment and the Economy*. Washington, D.C.: Island Press, 2001.

Parmesan, C., and G. Yohe. "A Globally Coherent Fingerprint of Climate-Change Impacts Across Natural Systems." *Nature* 421 (2003): 37–42.

Petschel-Held, G., A. Block, M. Cassel-Gintz, J. Kropp, M. K. B. Lüdeke, O. Moldenhauer, F. Reusswig, and H. J. Schellnhuber. "Syndromes of Global Change: A Qualitative Modeling Approach to Assist Global Environmental Management." *Environmental Modeling and Assessment* 4 (1999): 295–314.

Planchon, André. *La Saturation de la consommation.* Tours: Mame, 1974.

Ponting, Clive. *A Green History of the World: The Environment and the Collapse of Great Civilizations.* New York: St. Martin's Press, 1991.

Redman, Charles L. *Human Impact on Ancient Environments.* Tucson: University of Arizona Press, 1999.

Robertson, Roland. *Globalization: Social Theory and Global Culture.* London: Sage, 1992.

Root, T. L., J. T. Price, K. R. Hall, S. H. Schneider, C. Rosenzweig et al. "Fingerprints of Global Warming on Wild Animals and Plants." *Nature* 421 (2003): 57–60.

Schor, Juliet B. *The Overspent American: Upscaling, Downshifting, and the Consumer.* New York: Harper Perennial, 1998.

Seneca the Younger. *Naturales Quaestiones.* Loeb Library, ed. and trans. T. H. Corcoran. Cambridge, Mass.: Harvard University Press, 1971–72.

Trewavas, A. "Malthus Foiled Again and Again." *Nature* 418 (2002): 668–70.

Veblen, Thorstein. *The Theory of the Leisure Class.* New York: Macmillan, 1899.

Weber, Max. *Economy and Society: An Outline of Interpretive Sociology.* Edited by G. Roth and C. Wittich, translated by E. Fischoff et al. Berkeley: University of California Press, 1978.

Wernick, I. K. "Consuming Materials: The American Way." *Technological Forecasting and Social Change* 53 (1996): 111–22.

Wilk, R. "Consumption, Human Needs, and Global Environmental Change." *Global Environmental Change* 12 (2002): 5–13.

Young, Oran R. *The Institutional Dimensions of Environmental Change: Fit, Interplay, and Scale.* Cambridge, Mass.: MIT Press, 2002.

Chapter 5

Bourgenot, L. "Quelques réflexions sur l'histoire des forêts françaises." *Comptes rendus Académie agriculture français* 79 (1993): 85–92.

Braudel, Ferdinand. *L'Identité de la France: Espace et histoire.* Paris: Arthaud-Flammarion, 1986.

Clout, H. D. *The Land of France, 1815–1914.* London: Croom Helm, 1983.

Coe, Michael D. *The Maya.* 7th ed., revised and expanded. New York: Thames and Hudson, 2005.

Diamond, J. M. *Collapse: How Societies Choose to Fail or Succeed.* New York: Viking, 2005.

———. "Ecological Collapses of Past Civilizations." *Proceedings of the American Philosophical Society* 138 (1994): 363–70.

Gill, Richardson Benedict. *The Great Maya Droughts: Water, Life, and Death.* Albuquerque: University of New Mexico Press, 2000.

Haug, G. H., D. Günther, L. C. Peterson, D. M. Sigman, K. A. Hughen et al. "Climate and Collapse of Maya Civilization." *Science* 299 (2003): 1731–35.

Hodell, D. A., M. Brenner, J. M. Curtis, and T. Guilderson. "Solar Forcing of Drought Frequency in the Maya Lowlands." *Science* 292 (2001): 1367–70.

Hoyois, Giovanni. *L'Ardenne et l'Ardennais: L'évolution économique et sociale d'une région.* 2 vols. Gembloux: Duclot, 1949–53.

Leestmans, C. H. *Histoire d'une vallée: La Lienne en Haute-Ardenne, 1500–1800.* Stavelot: Chauveheid, 1980.

Mather, A. S. "The Transition from Deforestation to Reforestation in Europe." In *Agricultural Technologies and Tropical Deforestation*, edited by A. Angelsen and D. Kaimowitz, 35–52. New York: CABI/Center for International Forestry Research, 2001.

Mather, A. S., J. Fairbairn, and C. L. Needle. "The Course and Drivers of the Forest Transition: The Case of France." *Journal of Rural Studies* 15 (1999): 65–90.

Mather, A. S., and C. L. Needle. "The Forest Transition: A Theoretical Basis." *Area* 30 (1998): 117–124.

Mather, A. S., C. L. Needle, and J. R. Coull. "From Resources Crisis to Sustainability: The Forest Transition in Denmark." *International Journal of Sustainable Development and World Ecology* 5 (1998): 192–93.

Petit, C., and E. F. Lambin. "Long-Term Land-Cover Changes in the Belgian Ardennes (1775–1929): Model-Based Reconstruction vs. Historical Maps." *Global Change Biology* 8 (2002): 616–31.

Stuart, D. "Historical Inscriptions and the Maya Collapse." In *Lowland Maya Civilization in the Eighth Century A.D.*, edited by J. Sabloff and J. S. Henderson, 321–54. Washington, D.C.: Dumbarton Oaks, 1993.

Victor, D. G., and J. H. Ausubel. "Restoring the Forests." *Foreign Affairs* 79 (2000): 127–44.

Webster, David L. *The Fall of the Ancient Maya: Solving the Mystery of the Maya Collapse.* New York: Thames and Hudson, 2002.

Whitmore, T. M., B. L. Turner, D. L. Johnson, R. W. Kates, and T. R. Gottschang. "Long-Term Population Change." In *The Earth as Transformed by Human Action: Global and Regional Changes in the Biosphere Over the Past 300 Years*, edited by B. L. Turner et al. Cambridge: Cambridge University Press, 1990.

Chapter 6

Adger, W. N., T. A. Benjaminsen, K. Brown, and H. Svarstad. "Advancing a Political Ecology of Global Environmental Discourses." *Development and Change* 32 (2001): 681–715.

Behnke, R., I. Scoones, and C. Kerven, eds. *Range Ecology at Disequilibrium: New Models of Natural Variability and Pastoral Adaptation in African Savannas*. London: Overseas Development Institute, 1993.

Bovill, E. "The Encroachment of the Sahara on the Sudan." *Journal of the African Society* 20 (1920): 259–69.

Centre d'archives d'outre-mer, Affaires économiques, R24 (14 MI 1566). *Mission forestière.* Paris, 1907.

Ellis, J., and D. Swift. "Stability of African Pastoral Ecosystems: Alternate Paradigms and Implications for Development." *Journal of Rangeland Management* 41 (1998): 450–59.

Fairburn, W. *Report on Sylvan Conditions and Land Utilization in Northern Nigeria*. Kaduna: Ministry of Agriculture, 1937.

Fairhead, James, and Melissa Leach. *Misreading the African Landscape: Society and Ecology in a Forest-Savanna Mosaic*. Cambridge: Cambridge University Press, 1996.

Glantz, M. H. *Drought Follows the Plough: Cultivating Marginal Areas*. Cambridge: Cambridge University Press, 1994.

Gonzales, P. "Desertification and a Shift of Forest Species in the West African Sahel." *Climate Research* 17 (2001): 217–18.

Helldén, U. "Desertification: Time for a Reassessment?" *Ambio* 20 (1991): 372–83.

Hulme, M. "Climatic Perspectives on Sahelian Dessication: 1973–1998." *Global Environmental Change* 11 (2001): 19–29.

Hulme, M., and M. Kelly. "Exploring the Links between Desertification and Climate Change." *Environment* 35 (1993): 4–45.

Jouve, P. "Sécheresse au Sahel et stratégies paysannes." *Sécheresse* 2 (1991): 61–69.

Kerven, C., ed. *Prospects for Pastoralists in Kazakhstan and Turkmenistan: From State Farm to Private Flocks*. London: Curzon Press, 2001.

Lamprey, H. "Report on the Desert Encroachment Reconnaissance in Northern Sudan: 21 October to 10 November 1975." *Desertification Control Bulletin* 11 (1975): 1–7.

L'Hôte, Y., G. Mahé, B. Somé, and J.-P. Triboulet. "Analysis of a Sahelian Index from 1896 to 2000: The Drought Continues." *Hydrological Sciences Journal* 47 (2002): 563–72.

Lindqvist, S., and A. Tengberg. "New Evidence of Desertification from Case Studies in Northern Burkina Faso." *Geografiska Annaler* 75A (1993): 127–35.

Mabbut, J. A. "A New Global Assessment of the Status and Trends of Desertification." *Environmental Conservation* 11 (1984): 102–13.

Mainguet, Monique. *Desertification: Natural Background and Human Mismanagement*. Heidelberg: Springer-Verlag, 1991.

Mortimore, Michael, and William M. Adams. *Working the Sahel: Environment and Society in Northern Nigeria*. London: Routledge, 1999.

Nelson, R. *Dryland Management: The "Desertification" Problem*. World Bank Technical Paper 116. Washington, D.C.: International Bank for Reconstruction and Development, 1989.

Niamir-Fuller, M., ed. *Managing Mobility in African Rangelands: The Legitimization of Transhumance*. London: Intermediate Technology Publications, 1999.

Nicholson, S. E. "Long-Term Changes in African Rainfall." *Weather* 44 (1989): 46–56.

Niemeijer, D., and V. Mazzucato. "Soil Degradation in the West African Sahel: How Serious Is It?" *Environment* 44 (2002): 20–31.

Oba, G., N. C. Stenseth, and W. J. Lusigi. "New Perspectives on Sustainable Grazing Management in Arid Zones of Sub-Saharan Africa." *Bioscience* 50 (2000): 35–51.

Oldeman, L. R., R. T. A. Hakkeling, and W. G. Sombroek. *World Map on the Status of Human-Induced Soil Degradation*. Rev. ed. Wageningen, Netherlands: International

Soil Reference and Information Centre/Nairobi, Kenya: United Nations Environment Programme, 1991.

Prince, S. D., E. Brown de Colstoun, and L. L. Kravitz. "Evidence from Rain-Use Efficiencies Does Not Indicate Extensive Sahelian Desertification." *Global Change Biology* 4 (1998): 359–74.

Puigdefabregas, J. "Ecological Impacts of Global Change on Drylands and Their Implications for Desertification." *Land Degradation and Development* 9 (1998): 383–92.

Rasmussen, K., B. Fog, and J. E. Madsen. "Desertification in Reverse? Observations from Northern Burkina Faso." *Global Environmental Change* 11 (2001): 271–82.

Raynaut, Claude. *Societies and Nature in the Sahel.* London: Routledge, 1997.

Reynolds, J. F., and D. M. Stafford-Smith, eds. *Global Desertification: Do Humans Cause Deserts?* Berlin: Dahlem University Press, 2002.

Schlesinger, W. H., and N. Gramenopoulos. "Archival Photographs Show No Climate-Induced Changes in Woody Vegetation in the Sudan, 1943–1994." *Global Change Biology* 2 (1996): 137–41.

Stebbing, E. "The Encroaching Sahara: The Threat to West African Colonies." *Geographical Journal* 85 (1935): 506–24.

Tucker, C. J., H. E. Dregne, and W. W. Newcomb. "Expansion and Contraction of the Sahara Desert from 1980 to 1990." *Science* 253 (1991): 299–331.

United Nations Convention to Combat Desertification, Geneva, Switzerland. http://www.unccd.int.

Watts, M. J. "Social Theory and Environmental Degradation." In *Desert Development: Man and Technology in Sparse-Lands*, edited by Y. Gradus, 14–32. Dordrecht: D. Reidel, 1985.

Chapter 7

Arrhenius, S. "On the Influence of Carbonic Acid in the Air upon the Temperature of the Ground." *Philosophical Magazine* 41 (1896): 237–76.

Ausubel, J. H., and H. D. Langford, eds. *Technological Trajectories and the Human Environment.* Washington, D.C.: National Academy Press, 1997.

Ayres, Robert U. *Turning Point: An End to the Growth Paradigm.* London: Earthscan, 1998.

Cohen, J. E. "Population Growth and the Earth's Carrying Capacity." *Science* 269 (1995): 341–46.

Daily, Gretchen C., and Katherine Ellison. *The New Economy of Nature: The Quest to Make Conservation Profitable.* Washington, D.C.: Island Press, 2002.

Dobson, A. P., A. D. Bradshaw, and A. J. M. Baker. "Hopes for the Future: Restoration Ecology and Conservation Biology." *Science* 277 (1997): 515–22.

Elgin, Duane. *Voluntary Simplicity: Toward a Way of Life That Is Outwardly Simple, Inwardly Rich.* New York: Quill, 1998.

Gleick, P. H. "Soft Water Paths." *Nature* 418 (2002): 373.

Grübler, Arnulf. *Technology and Global Change.* Cambridge: Cambridge University Press, 1998.

Hawken, Paul, Amory Lovins, and L. Hunter Lovins. *Natural Capitalism: Creating the Next Industrial Revolution*. Boston: Little, Brown, 1999.

Hoffert, M. I., K. Caldeira, G. Benford et al. "Advanced Technology Paths to Global Climate Stability: Energy for a Greenhouse Planet." *Science* 298 (2002): 981–87.

Hutter, B. "Compliance and Beyond." *European Business Forum*. Special issue on Sustainable Development (2002): 11–12.

Kates, R. W. "Population and Consumption: What We Know, What We Need to Know." *Environment* 42 (2000): 10–19.

Keith, D. W. "Geoengineering the Climate: History and Prospect." *Annual Review in Energy and the Environment* 25 (2000): 245–84.

Larssen, S., K. J. Barrett, J. Fiala, J. Goodwin, L. O. Hagen et al. *Air Quality in Europe: State and Trends, 1990–1999*. Working Paper 4. Copenhagen: European Environment Agency Working Paper, 2002.

Liu, J., G. C. Daily, P. R. Ehrlich, and G. W. Luck. "Effects of Household Dynamics on Resource Consumption and Biodiversity." *Nature* 421 (2003): 530–33.

Lutz, W., W. Sanderson, and S. Scherbov. "The End of World Population Growth." *Nature* 412 (2001): 543–45.

McNeill, J. R. *Something New Under the Sun: An Environmental History of the Twentieth-Century World*. New York: W. W. Norton, 2001.

Myers, Norman, and Jennifer Kent. *Perverse Subsidies: How Tax Dollars Can Undercut the Environment and the Economy*. Washington, D.C.: Island Press, 2001.

Norgaard, Richard B. *Development Betrayed: The End of Progress and a Coevolutionary Revisioning of the Future*. London: Routledge, 1994.

Orr, David W. *The Nature of Design: Ecology, Culture, and Human Intention*. New York: Oxford University Press, 2002.

Princen, T., M. Maniates, and K. Conca, eds. *Confronting Consumption*. Cambridge, Mass.: MIT Press, 2002.

Rosenblatt, R., ed. *Consuming Desires: Consumption, Culture, and the Pursuit of Happiness*. Washington, D.C.: Island Press, 1999.

Schor, Juliet B. *The Overspent American: Upscaling, Downshifting, and the Consumer*. New York: Harper Perennial, 1998.

Schumacher, E. F. *Small Is Beautiful: Economics as if People Mattered*. Vancouver: Hartley and Marks, 1973.

Segal, Jerome M. *Graceful Simplicity: The Philosophy and Politics of the Alternative American Dream*. Berkeley: University of California Press, 2003.

Smil, Vaclav. *Feeding the World: A Challenge for the Twenty-First Century*. Cambridge, Mass.: MIT Press, 2000.

Stahel, Walter R., and Max Börlin. *Stratégie économique de la durabilité*. Geneva: Société de la Banque Suisse, 1987.

Wernick, I. K. "Consuming Materials: The American Way." *Technological Forecasting and Social Change* 53 (1996): 111–22.

World Population Prospects: The 2002 Revision. Working Paper 180, I–IX. Washington, D.C.: The United Nations Population Division, 2003.

Conclusion

Berkes, F., J. Colding, and C. Folke, eds. *Navigating Social-Ecological Systems: Building Resilience for Complexity and Change.* Cambridge: Cambridge University Press, 2003.

Havel, V. "Les flèches du renouveau." Published worldwide in daily newspapers by the Project Syndicate, 2000.

Landes, David. *The Wealth and Poverty of Nations: Why Some Are So Rich and Some So Poor.* New York: W. W. Norton, 1998.

Schor, J. B., and B. Taylor, eds. *Sustainable Planet: Solutions for the Twenty-First Century.* Boston: Beacon Press, 2003.

Wilbanks, T. J. "Sustainable Development in Geographic Perspective." *Annals of the Association of American Geographers* 84 (1994): 541–56.